Statistical Prediction and Machine Learning

Written by an experienced statistics educator and two data scientists, this book unifies conventional statistical thinking and contemporary machine learning framework into a single overarching umbrella over data science. The book is designed to bridge the knowledge gap between conventional statistics and machine learning. It provides an accessible approach for readers with a basic statistics background to develop a mastery of machine learning. The book starts with elucidating examples in Chapter 1 and fundamentals on refined optimization in Chapter 2, which are followed by common supervised learning methods such as regressions, classification, support vector machines, tree algorithms, and range regressions. After a discussion on unsupervised learning methods, it includes a chapter on unsupervised learning and a chapter on statistical learning with data sequentially or simultaneously from multiple resources.

One of the distinct features of this book is the comprehensive coverage of the topics in statistical learning and medical applications. It summarizes the authors' teaching, research, and consulting experience in which they use data analytics. The illustrating examples and accompanying materials heavily emphasize understanding on data analysis, producing accurate interpretations, and discovering hidden assumptions associated with various methods.

Key Features:
- Unifies conventional model-based framework and contemporary data-driven methods into a single overarching umbrella over data science.
- Includes real-life medical applications in hypertension, stroke, diabetes, thrombolysis, aspirin efficacy.
- Integrates statistical theory with machine learning algorithms.
- Includes potential methodological developments in data science.

John Tuhao Chen is a professor of Statistics at Bowling Green State University. He completed his postdoctoral training at McMaster University (Canada) after earning a PhD degree in statistics at the University of Sydney (Australia). John has published research papers in statistics journals such as Biometrika as well as in medicine journals such as the Annals of Neurology.

Lincy Y. Chen is a data scientist at JP Morgan Chase & Co. She graduated from Cornell University, winning the Edward M. Snyder Prize in Statistics. Lincy has published papers regarding refinements of machine learning methods.

Clement Lee is a data scientist in a private firm in New York. He earned a Master's degree in applied mathematics from New York University, after graduating from Princeton University in computer science. Clement enjoys spending time with his beloved wife Belinda and their son Pascal.

Statistical Prediction and Machine Learning

John Tuhao Chen
Lincy Y. Chen
Clement Lee

CRC Press
Taylor & Francis Group
Boca Raton London New York

CRC Press is an imprint of the
Taylor & Francis Group, an **informa** business

A CHAPMAN & HALL BOOK

Designed cover image: © John Tuhao Chen

First edition published 2024
by CRC Press
2385 NW Executive Center Drive, Suite 320, Boca Raton FL 33431

and by CRC Press
4 Park Square, Milton Park, Abingdon, Oxon, OX14 4RN

CRC Press is an imprint of Taylor & Francis Group, LLC

© 2024 Taylor & Francis Group, LLC

ISBN: 978-0-367-33227-3 (hbk)
ISBN: 978-1-032-81272-4 (pbk)
ISBN: 978-0-429-31868-9 (ebk)

DOI: 10.1201/9780429318689

Typeset in Nimbus Roman
by KnowledgeWorks Global Ltd.

Publisher's note: This book has been prepared from camera-ready copy provided by the authors.

To the memory of my parents
HongBiao Chen and WanJuan Lin

Contents

Preface

Big data has penetrated into every corner of our lives. Its omnipresence and the demands of the market necessitates that we decently understand it. To this end, new books and literature that explain techniques in data-oriented prediction and machine learning are required. Although there are excellent textbooks focusing on data analysis with conventional statistical approaches, as well as outstanding textbooks addressing machine learning methods for data-oriented approaches, to this date, nothing has merged the two comprehensively. These two approaches over time have led to two primary camps in data science, one focused on data-oriented analysis and another on model-based analysis. Students, data analysts, and junior researchers are often confused about which camp they may fall under, especially as the two data analytic camps are often seemingly contradictory to each other. There is much debate on the right camp to select in the broader realm of data science. Written by an experienced statistician and two data scientists, this book unifies the two frameworks into a single overarching umbrella on data science.

Starting from a background in a basic undergraduate college statistics course, the conventional model-based inference framework finds its foundations in data analytics. It consists of an underlying model for the data, hypothesis testing or confidence estimation on unknown model parameters, measuring variations behind the data, and prediction an unknown quantity related to the inference problem. Under this style of thinking, the underlying model serves as the hub in data analysis. An implausible model assumption may thus result in *a correct answer towards the wrong problem*, which can often lead to misleading prediction results.

When addressing practical problems such as high dimensional inference, machine learning often relies on computer intensive algorithms. Many of the underlying thought processes and methodologies have been well-developed but are still fundamentally based in the conventional data analysis framework. One of the major challenges underpinning modern machine learning stems from the gap between the conventional model-based inference and data-driven learning algorithms. The knowledge gap hinders practitioners (especially students, researchers, data analysts, or consultants) from truly mastering and correctly applying machine learning skills in data science.

This book is designed to bridge the gap between conventional statistics and machine learning. It provides an accessible approach for readers with a basic statistical background to develop a mastery of machine learning. We start with elucidating examples in Chapter 1 and introducing fundamentals on refined optimization in Chapter 2, which are followed by common supervised learning methods such as regressions, classification, support vector machines, tree algorithms, and range regressions. After a discussion on unsupervised learning methods, we include a chapter on unsupervised learning as well as a chapter on statistical learning with data sequentially or simultaneously from multiple resources.

One of the distinct features of this book is the comprehensive coverage of the topics in statistical learning. This book summarizes the authors' teaching, research, and consulting experience in which they used data analytics. The illustrating examples and accompanying materials heavily emphasize understanding on the two camps described above, producing accurate interpretations, and discovering hidden assumptions associated with various data analysis methods. It is designed to guide students toward effectively applying statistical learning methods.

This book is addressed to practitioners in data science, but it is also suitable for upper-level undergraduate students and entry-level graduate students who are interested in obtaining a more thorough comprehension of machine learning. The potential audience extends to data scientists who are interested in more insightful interpretations of raw outputs generated from machine learning. The materials of the book originate from the first author's lecture notes of a one-semester machine learning course taught at the University of California Berkeley.

We are grateful to a number of people who have encouraged and contributed to the writing of this book. Thanks to Professor Deb Nolan at UC Berkeley, whose conversations and opinions partially motivated the writing plan. The writing of the book also benefitted from suggestions, questions, and clarifications of the students of Berkeley Stat 154 (Fall 2018) class, especially Haotian Fu who developed an R package for range regression. We also thank the following PhD and Master students at BGSU for their assistance and contributions, Gul Bulbul, Asmita Ghoshal, Chao Gu, Yiheng Liu, Rachana Mahajan, Corey Thrush, and Peiyao Wang.

We owe a great deal to our families for their great support, especially to Binglin for her numerous suggestions; Belinda for being the supreme leader of everything related to the book writing; Vincent and Janet for helpful discussions; and Patrick for his tireless reading and editing.

The reviewers' comments and suggestions have also significantly improved the original writing plan of this book. Their contributions are gratefully acknowledged. Of course, we are solely responsible for unavoidable typos and errors. Finally, we would like to thank this book's acquisitions editor, Mr. David Grubbs, and editorial assistant, Mr. Curtis Hill, at Chapman and Hall/CRC, for their helpful efforts, kindness, and patience (especially allowing us to continuously postpone the final manuscript submission from one weekend to another weekend) during this project.

List of Figures

List of Tables

1

Two Cultures in Data Science

Data is a starting point in model prediction as well as in machine learning addressed in the celebrated books on statistical learning [56], [72]. With a set of data, data scientists predict future responses with certain level of scientific reliability and confidence on the accuracy of prediction. In the process of data analysis, for example, one way is to use linear regression model in conjunction with normality assumption on the distribution of data. Another way is to use tree regressions or neural network without any model assumption. As elucidated by Breiman in [13] and [11], one culture of prediction is to start with an assumed relationship that underpins the response and the explanatory variables in conjunction with a random model that governs the distribution of the data fluctuation. Data scientists then follow up with inference issues such as coefficient estimation, hypothesis testing, and prediction. This is the model-based prediction approach. Another culture of prediction is to start with the data without making any model assumption. It uses training data to build a model, and make predictions based on the trained model. This is commonly referred to as a data-driven approach. In practice, data scientists, shaped by their believes and operating mechanism in data analytics, intentionally (or unintentionally) fall into either the model-based culture camp or the data-driven culture camp in prediction. In this chapter, we will discuss intrinsic connections, compare evaluation criteria, and address optimality issues related to the two cultures.

Generally, when the sample size is small or moderate, model-based approach is able to accommodate additional model information on the data to alleviate difficulties caused by insufficient sample sizes. On the other hand, when the sample size is reasonably large, the data-based approach has the advantage of avoiding implausible model assumptions and digging out the underlying knowledge hidden behind the data.

1.1 Model-based culture

The culture camp of model-based inference mainly consists of statisticians who start data analysis with the assumption on the model governing the random mechanism of the data. This includes the model-based inference in which the model underlying the data is explicitly or implicitly assumed. It also includes non-parametric statistical analysis in which the inference is motivated by and grounded on a set of general population homogeneity (such as common continuous cumulative distribution functions). The model-based approach has been well documented in conventional statistical analyses, for instance, [9], [10], [43], and [91], among others. In this approach, we assume that the data set is generated from a population with unknown parameters:

$$y = f(\mathbf{x}|\eta) + \epsilon,$$

where y is the response, $f(\mathbf{x}|\eta)$ is a specific function with unknown parameters η, and the random fluctuation is denoted by ϵ. The question of interest is "how to use a set of data to estimate or perform hypothesis testing on the unknown parameters η for prediction?". In practice, the format of $f(\mathbf{x}|\eta)$ is usually dictated by the nature of the problem, such as the linear models for continuous responses, logistic regression models for binary responses, or log linear models for skewed responses, to list just a few. To further illustrate this point, we discuss the following example in statistical prediction.

Suppose that we observe a quantitative response Y and p different predictors $X_1, ..., X_p$, with the model,

$$Y = f(X_1, ..., X_p) + \epsilon, \tag{1.1}$$

where $f(.)$ is an unknown function that contains systematic information encoded in the predictors for the response variable, and ϵ is the random error term.

To seek the estimation on the unknown function $f(.)$, $\hat{f}(.)$, we split the original data into a training set (usually 75% of the original data) to make inference on the unknown function $f(.)$, with the aim of minimizing the mean prediction error. In this regard, statistical learning refers to a set of approaches for making inference about the unknown function $f(.)$ in (1.1).

The following two examples explain the selection of $f(.)$ in model-based inference.

Example 1.1 *Consider the hypothetical data on lung cancer patients in a clinical trial listed in Table 1.1. To analyze possible associations between tentative risk factors and lung cancer, a logistic regression model*

$$log(\frac{p}{1 - p}) = \alpha + \mathbf{x}'\beta,$$

points to the underlying connection

$$P(Y = 1) = \frac{e^{\alpha+\mathbf{x}'\beta}}{1 + e^{\alpha+\mathbf{x}'\beta}}.$$

In Example 1.1, the response variable is the positive clinical outcome in lung cancer, while the predictors may include gender, age, smoking, diabetes, and

TABLE 1.1

Lung cancer data structure

Item	Patient#1	Patient#2	Patient#3	Patient#4
ID	N01101	N01102	N01103	N01104
Gender	M	F	M	F
Age	30	43	71	63
Smoking	Y	N	Y	Y
Diabetes	N	Y	Y	N
Hypertension	Y	N	Y	Y
Lung Cancer	Y	N	N	Y

hypertension. As shown in the data types in Table 1.1, the usual normality assumption fails, thus it is not appropriate to use the linear regression model to fit the data. Instead, since the type of data is case-control data, a logistic regression model would be more appropriate to analyze the odds ratio on the disease rate of lung cancer associated with the population defined by the strata related to the combination of risk factors. The case-control feature of the data determines the analytical approach on the unknown function $f(.)$ and the prediction outcome on severity of risk factors associated with the disease.

The discrete feature of the response variable in Example 1.1 determines the logistic function for the underlying model $f(.)$ because the outcome of developing lung cancer is either "yes" or "no". The next example takes the approach of simple linear regression since the response variable Y, insurance premium, is continuous. It sets the connection between insurance premium and driving experience, and demonstrates that within the method of simple linear regression, model-based prediction discerns greatly from data-driven prediction in the learning process toward the underlying model $f(x|\eta)$.

Example 1.2 *Assume that the insurance premium linearly decreases as the driving year increases, more specifically, we assume the model behind the data as*

$$y = \alpha + \beta x + \epsilon,$$

where y is the insurance premium, x is the driving experience in years, α is the intercept for the mean premium of a new driver who has no driving experience, and β is the slope for the amount of decrease in monthly insurance premium for the increase of each driving year. The error term ϵ is the random variation attributable to other factors such as age, gender, income, marital status, etc. The learning process toward $f(.)$ is tantamount to the estimation of model parameters α and β.

As usual, assume that ϵ follows a normal model with an unknown standard deviation σ. In regression analysis, we estimate the values of the parameters α and β, and use the estimated model

$$\hat{y}(x) = \hat{\alpha} + \hat{\beta}x$$

for prediction in Example 1.2.

Now, consider the relationship between 6-month insurance premium and driving experience for the data set insurance.txt. If we assume that the insurance premium (y) decreases as the driving experience (x) increases, and fit a linear regression model, we essentially assume that

$$y = \alpha + \beta x + \epsilon.$$

```
Call:
lm(formula = x[, 2] ~ x[, 1])

Residuals:
    Min      1Q  Median      3Q     Max
-172.281 -44.548   1.032  48.874 153.402

Coefficients:
             Estimate Std. Error t value Pr(>|t|)
(Intercept) 544.6177     2.6766  203.47   <2e-16 ***
x[, 1]      -23.8851     0.3639  -65.63   <2e-16 ***
---
Signif. codes:  0 '***' 0.001 '**' 0.01 '*' 0.05 '.' 0.1 ' ' 1

Residual standard error: 52.83 on 1249 degrees of freedom
Multiple R-squared:  0.7752,  Adjusted R-squared:  0.775
F-statistic:  4308 on 1 and 1249 DF,  p-value: < 2.2e-16
```

FIGURE 1.1
Premium-time regression analysis of the insurance data

As shown in Figure 1.1, the estimated intercept in the linear regression model is $\hat{\alpha} = 544.62$, which means that the long-term average of 6-month insurance premium for a new driver is \$544.62 (since driving experience = 0). The estimated regression coefficient is $\hat{\beta} = 23.89$, which means that for each additional year of driving, the 6-month insurance premium decreases, on average, \$23.89. Both p-values for the intercept and regression coefficient are less than 0.0001, suggesting that the model is statistically significant.

The validity of the above analysis is basically grounded on the following two assumptions. First, the relationship between the 6-month insurance premium and the driving experience is a linear function,

$$f(x) = \alpha + \beta x.$$

Second, the random fluctuation term ϵ follows a normal model. If any one of these assumptions fails, the corresponding data analysis becomes invalid and the conclusion would be misleading. In the application of real-life data analyses, the plausibility of the model assumptions usually comes from the information of the data in the specific field of investigation. However, in situations where informative knowledge is not available for the assumption of the specific form of $f(\mathbf{x}|\eta)$, carelessly applying a linear model (or a generalized linear model) to a set of data may result in misleading conclusions, albeit the model may be statistically significant.

In the next section, we follow up on the analysis of the insurance data in Example 1.2 to illustrate that the model-based analyses, especially the simple linear structure of $f(.)$ in Example 1.2 may be actually wrong, as discussed in data-driven analyses in the next section.

1.2 Data-driven culture

The data-driven culture camp is basically grounded on the belief that the data contains all information needed for prediction, without any additional model assumption. Representing approaches in this culture camp include unsupervised machine learning, decision trees, classification, range regression, neural network, and deep learning, to list just a few.

As discussed in the previous section, when we make statistical inference using model-based approach, the plausibility of the model assumptions is critical. Invalidity on one of the model assumptions may consequently ruin the whole data analysis. And the analysis becomes a "correct answer to a wrong problem" as commented by Tukey in 1962 (See, for instance, [11],[12], and [118]). When the underlying model of the data cannot be plausibly and legitimately assumed, one way to approach the correct model, $f(.)$, when the data is large enough is the data-driven approach. Self-evidently in the sequel, we also use the term "data-based approach" or "data-oriented approach" to refer to the same approach in seeking the underlying model behind the data. The data-driven approach usually starts with the plot of the insurance premium data, and use the pattern of the data to study the plausibility of the model assumptions.

We shall use the insurance premium example, Example 1.2, that we discussed with the model-based approach in the preceding section.

Example 1.3 *The premium-year plot in Figure 1.2 indicates that the insurance premium is not a linear function of driving year.*

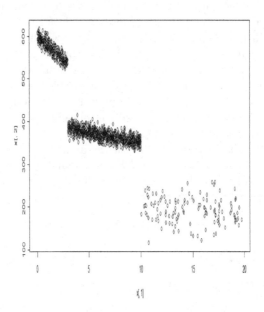

FIGURE 1.2
Premium-year plot of the insurance data

As shown in Figure 1.2, there is actually no linear pattern for the relationship between the 6-month insurance premium and the driving year. What is behind the data is more likely a piece-wise linear function. Figure 1.2 depicts that during the first 3 years, the decrease of the insurance premium for each year of driving is much larger than the corresponding change for customers who have driven 3 to 10 years. After 10 years, essentially there is no gain in insurance premium for any additional year of driving. This is more close to the realistic practice in the way that new drivers are charged with higher rates for the first few years (the first stage). In the second stage (3 to 10 years), although the rate decreases as the driving year increases after the initial stage (0-3 years), the slope is relatively more stable compared with the changes in the first stage. After 10 years driving, drivers essentially get a flat rate that has nothing related to an additional year of driving.

In what follows, we shall examine the statistical significance of the three phases separately.

Figure 1.4 provides the outcome of linear regression analysis for the effect of driving years on the 6-month insurance premium. For drivers with less

FIGURE 1.3
Plausibility of piece-wise function for insurance premium

than three years' experience, the long-term average rate is $601.34 (when the driving experience $x = 0$). And for each additional year of driving, the 6-month premium, on average, drops $21.62 within the first three years of driving. The two coefficients are significantly different from zero because the p-values are much less than 0.05.

The regression result in Figure 1.5 indicates that people with three to five years driving experience are charged at a basic rate, on average, $399.36. Such a rate decreases by $4.85, on average, for each additional year of driving. Compared with drivers in the first three years of driving, this group of drivers pays a lower starting premium ($601.34 versus $399.36), but the rate of change for each additional year of driving is much less ($21.62 versus $4.85). All the estimated parameters are statistically significant.

For the third consumer group who have driven more than 10 years, as shown in Figure 1.6, people with more than 10 years driving experience basically pay an average flat rate of $211.01 (statistically significant) in which the impact of an additional year of driving on the 6-month premium is not statistically significant (the corresponding p-value is 0.169).

The above example indicates that the linear model as in Figure 1.1 can be completely different from the true model behind the data even through the p-value of the model is statistically significant. Starting with a correct model assumption is critical in data analytics. If the initial model assumption

```
> summary(lm.fit1)

Call:
lm(formula = x1[, 2] ~ x1[, 1])

Residuals:
     Min      1Q  Median      3Q     Max
 -25.9759 -6.5179 -0.5958  6.5865 27.1037

Coefficients:
             Estimate Std. Error t value Pr(>|t|)
(Intercept) 601.3411     1.1606  518.12   <2e-16 ***
x1[, 1]     -21.6200     0.6917  -31.26   <2e-16 ***
---
Signif. codes:  0 '***' 0.001 '**' 0.01 '*' 0.05 '.' 0.1 ' ' 1

Residual standard error: 10.24 on 300 degrees of freedom
Multiple R-squared:  0.7651,  Adjusted R-squared:  0.7643
F-statistic:   977 on 1 and 300 DF,  p-value: < 2.2e-16
```

FIGURE 1.4
Premium-time regression for inexperienced drivers

is wrong, we may end up answering a question completely different from the
goal of the data analysis.

In general, the data-driven approach starts with

$$y = g(\mathbf{x}) + \epsilon,$$

where $g(\mathbf{x})$ here is the function governing the response y, and ϵ is the random
variation of the data. In data-driven inference, we do not assume any specific
function of $g(\mathbf{x})$. Instead, we plot the data and use the pattern shown in the
diagram to *learn* the shape of $g(\mathbf{x})$, denoted as $\hat{g}(\mathbf{x})$.

Specifically, in the analysis of the insurance premium data, instead of starting with the conventional linear regression model, after plotting the data in
Figure 1.2, we are able to *regularize* the piece-wise function which reflects
the general premium-decision policy of the company. This special function
$g(.)$ actually governs the general relationship between insurance premium and
driving years.

Data-driven approach usually separates the original data into two portions.
One is the training set to learn about $\hat{g}(x)$, and another is the testing set to

```
> summary(lm.fit2)

Call:
lm(formula = x2[, 2] ~ x2[, 1])

Residuals:
    Min      1Q  Median      3Q     Max
-28.652  -6.856  -0.190   6.670  34.447

Coefficients:
             Estimate Std. Error t value Pr(>|t|)
(Intercept) 399.3644     1.1719   340.8   <2e-16 ***
x2[, 1]      -4.8516     0.1733   -28.0   <2e-16 ***
---
Signif. codes:  0 '***' 0.001 '**' 0.01 '*' 0.05 '.' 0.1 ' ' 1

Residual standard error: 10.04 on 819 degrees of freedom
Multiple R-squared:  0.4891,   Adjusted R-squared:  0.4884
F-statistic: 783.9 on 1 and 819 DF,  p-value: < 2.2e-16
```

FIGURE 1.5
Premium-time regression for drivers with 3-10 years of experience.

evaluate the trained outcome. Theoretically, as pointed out in Breiman and Friedman [14], or Stone [113], one of the general principles or criteria for the selection of the underlying function behind the data is the EPE (expected prediction error, or expected squared prediction error).

Definition 1.1 EPE: *Let Y be the response observation and $\hat{Y}$ be the prediction of the response based on a set of data, the expected squared prediction error (EPE) is defined as*

$$EPE = E[(Y - \hat{Y})^2].$$

Note that in the above definition, if $\hat{Y}$ is a predicted value of Y based on a set of training data, and $\{(x_i, y_i), i = 1, ..., k\}$ is a set of test data, the estimate of the EPE is

$$\hat{EPE} = \frac{1}{k} \sum_{i=1}^{k} (y_i - \hat{y}_i)^2,$$

where for any $i = 1, ..., k$, the predicted response is the value of the corre-

```
> summary(lm.fit3)

Call:
lm(formula = x3[, 2] ~ x3[, 1])

Residuals:
    Min     1Q  Median     3Q     Max
-81.396 -17.107   0.441  19.073  68.938

Coefficients:
             Estimate Std. Error t value Pr(>|t|)
(Intercept)  211.010     13.807   15.283   <2e-16 ***
x3[, 1]       -1.264      0.913   -1.385    0.169
---
Signif. codes:  0 '***' 0.001 '**' 0.01 '*' 0.05 '.' 0.1 ' ' 1

Residual standard error: 29.74 on 126 degrees of freedom
Multiple R-squared:  0.01499,  Adjusted R-squared:  0.007169
F-statistic: 1.917 on 1 and 126 DF,  p-value: 0.1686
```

FIGURE 1.6
Premium-time regression for drivers with more than 10 years experience

sponding trained model $\hat{g}_i$,

$$\hat{y}_i = \hat{g}_i(x_1, ..., x_k).$$

It should be noted that when we apply the mean (squared) prediction error as a criterion to evaluate the trained function $\hat{g}(\mathbf{x})$, the optimal solution $\hat{g}(\mathbf{x})$ takes the form $E(Y|\mathbf{X})$, as shown in the following theorem.

Theorem 1.1 *Let Δ be a set of permissible functions of $g(\mathbf{x})$ for the model $Y = g(\mathbf{x}) + \epsilon$, we have*

$$argMin_{g\in\Delta}E_{Y|X}[|Y - g(\mathbf{x})|^2|\mathbf{X} = \mathbf{x}] = E(Y|\mathbf{X} = \mathbf{x}).$$

Theorem 1.1 indicates that the model minimizing the expected prediction error is the conditional expected value of the response given the features associated to the response of interest. Details on the proof of this theorem can be found in [56], [119], or [120]. We will also discuss this result in Section 1.4.1 when we discuss the outcome evaluation in model-based inference.

1.3 Intrinsics between the two culture camps

It should be noted that the two culture camps of data science have their own advantage and disadvantage. For instance, the model-based camp may cooperate implausible assumptions into the model, while the data-driven approach may over-fit the model with random features contained in the data. Both ways of data analysis result in errors in prediction. This necessitates a discussion on the criterion measuring the fitness of the learned model to the data.

From the insurance premium examples discussed in the preceding sections, it seems that the data-driven approach leads to a closer description of the unknown underlying relationship between the response and the features (predictors). It avoids implausibly making model assumptions before examining the data. However, this is true only when the size of the data is large enough to unveil the underlying relationship, and the random features contained in the data are not over-fitted into the model. If the sample size is not large enough, the plot of the data may show patterns merely pertaining to the training data. This may mislead the prediction in the form of the true underlying relationship $f(\mathbf{x})$. In this case, the additional information on the model helps to navigate toward a more legitimate conclusion.

1.3.1 Small sample inference necessitates model assumptions

Consider a set of simulated data from a normal model with small sample size and large standard deviation. Since the sample size is small, if we fit the underlying distribution using the simulated data, we may get into a completely different model. As shown in Figure 1.7, although the dataset was originally generated from $N(0, 7)$, since the sample size is only 15, the p-value is 0.0001152 rejecting the hypothesis that the data is from a normal model. Instead, the dataset fits well with an exponential model (p=0.1463) using one-sample Kolmogorov-Smirnov test.

This example indicates that when the sample size is not large enough, there is a risk of obtaining a misleading result from data-driven approach. Under this scenario, additional information including appropriate model assumptions, such as assuming that the underlying model is skew normal, or a regular normal model $N(\mu, \sigma^2)$ with unknown parameters μ and σ, may help to regularize the estimation of the underlying model toward the right direction.

It is also related at this point to note that the non-parametric statistical method actually pertains to the model-based culture camp. Regarding the difference between model-based approach and data-driven approach in data science, one of the confusing issues is the methodology named "distribution-free" or "model-free" statistical methods. In fact, the "distribution-free" in-

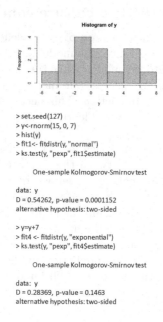

```
> set.seed(127)
> y<-rnorm(15, 0, 7)
> hist(y)
> fit1<- fitdistr(y, "normal")
> ks.test(y, "pexp", fit1$estimate)

        One-sample Kolmogorov-Smirnov test

data: y
D = 0.54262, p-value = 0.0001152
alternative hypothesis: two-sided

> y=y+7
> fit4 <- fitdistr(y, "exponential")
> ks.test(y, "pexp", fit4$estimate)

        One-sample Kolmogorov-Smirnov test

data: y
D = 0.28369, p-value = 0.1463
alternative hypothesis: two-sided
```

FIGURE 1.7
Small dataset misleads the underlying model

ference approach, as named in non-parametric statistics essentially goes with assumptions on the underlying model for the data. Thus, it is actually a model-driven inference method. For instance, the distribution-free Wilcoxon signed rank statistic is based on the assumption that the underlying model can be characterized by the median of the population, and the underlying distribution is symmetric around the population median. As delineated by Hollander and Wolfe [59] as well as Sprent and Smeeton [112], the terminology of "distribution-free" refers to the condition that the model assumption does not involve an explicit form of the underlying model. It still contains assumptions on the general shape for inference, focusing on a class of model features. This is unlike the data-driven techniques, in which the there is completely no assumption on the underlying model. Thus, the distribution-free non-parametric statistics is classified as a model-oriented inference approach, not data-oriented inference approach.

1.3.2 Prediction accuracy demands large sample

One of the drawbacks on model-based approach is that almost all the model assumptions are oversimplified for real-life data due to the complexity of real-life situations. When a data set with a reasonable sample size is available, it is more legitimate to dig out the pattern (model information) directly from the data. The advantage of data-driven approach, when available, is to avoid blind model assumption on $f(.)$, thus it is more reliable in prediction. Studies of data-oriented approaches constitute the major part of data science. Certainly, this way of establishing a model is essentially different from the classical statistics in which we start with a set of random samples from a population before making prediction. Since the methodology only valid when the sample size is large enough, it should be noted that practices of data-driven approaches with small sample are usually misleading and risky.

It is related at this point to emphasize that the data-driven (or data-oriented) approach is grounded upon asymptotic properties of functions of the available data. The accuracy and reliability of the data-oriented approach heavily rely on the available data. That is part of the well-known drawback of "data-hungry" issue in data-driven model analysis.

1.3.3 Which camp to go?

Theoretically, it seems that the sample size is a key component in the selection between model-based analytics and data-oriented analytics. However, for a set of data, there is no unified way to determine how large is large enough for the sample size for the selection of inference approaches. Similarly, without checking the background information of the data, it is risky to assert appropriateness on the model-based approach or the data-oriented approach, . In practice, some rules of thumb are commonly applicable. For instance, the background information of the data may permit a plausible model assumption for a function of some existing models (even though it is approximate), This includes a linear combination of mix normal models for the height of a population, or a combination of mix exponential models for time effects. See for example [84], [42], and [40]. On the other hand, if the information on the distribution of the data is completely unknown, data-driven methods such as the spline methods or decision trees are more appropriate for the analysis (see, for example, [13], [120], and [121], to list just a few).

In the case where no distinguishable features of the data set are available to justify whether we should go with the model-based camp or data-driven camp, the performance of the approach becomes the follow-up criterion for consideration. Inference methods with better performance should be selected. For instance, the model-based camp evaluates the performance of inference approach by expected prediction error, controls the type-I error with high power in hypothesis testing of the associated model parameters, or high confidence level and low estimate error in the estimation. On the other hand, the data-driven culture camp uses testing errors, such as false positive rate, false negative rate, or expected mean prediction error computed with the testing data. For example, in a prediction problem, generally, the method that predicts more closely to the true value should be selected.

As discussed above, when the dataset does not contain adequate information on the selection of inference culture camp, the evaluation criterion and accuracy measurement essentially dictator the selection of prediction methods between model-based camp or data-driven camp. However, each camp has its own well-developed criteria on measuring closeness for various inference problems. In what follows, we shall discuss the evaluation criteria and optimizing strategies on the evaluation of inference performance for each of the two analytic culture camps in data science. Specific concepts and terminologies pertaining to a specific algorithm will be addressed in the corresponding chapter when the topic is discussed.

1.4 Learning outcome evaluation

This section contains discussions on cases of evaluation criteria for model-driven analytic approach and data-driven approach. We shall start with model-based approaches. With rapid development of statistical theory today, it is impossible to exhaust all methods of model evaluation in one section. As an introduction, we will use hypothesis testing as an illustrating example to elucidate the evaluation criteria in model-based inference. Hypothesis testing is a common approach covered in almost every statistics textbook.

Following the discussion on model-based approach, we will discuss logistic regression built from a set of training data in the next subsection to discuss evaluation criteria for data-oriented analytic approaches.

1.4.1 Error rates in model-based culture camp

One of the common evaluation criteria is the expected prediction error (EPE) which directly measures the squared error between the observed and the predicted values.

$$EPE(f) = E(Y - f(X))^2$$
$$= \int [y - f(x)]^2 dF(x, y) \tag{1.2}$$

Since X is the predictor, by conditioning on X, we have

$$EPE(f) = E_X\{E_{Y|X}([Y - f(X)]^2 | X)\}.$$

Thus, the search for the underlying function $f(x)$ can be achieved by minimizing the EPE point-wise.

$$f(x) = argMin_c E_{Y|X}([Y - c]^2|X = x).$$

Now, consider

$$
\begin{aligned}
&E_{Y|X}([Y - c]^2|X = x) \\
&= E_{Y|X}[(Y - E(Y|X = x) + E(Y|X = x) - c)^2|X = x] \\
&= E_{Y|X}\{[Y - E(Y|X = x)]^2|X = x\} + E_{Y|X}\{[E(Y|X = x) - c]^2|X = x\}
\end{aligned}
$$

Thus, when the value c takes $E(Y|X = x)$, the conditional expected value reaches its smallest possible value. And the solution is

$$f(x) = E(Y|X = x).$$

To illustrate how the model-based approach evaluates the performance of an inference procedure, we discuss a simple example in hypothesis testing on the inference for a normal population mean. In this case, the evaluation criterion is the power and significance level of the test.

Example 1.4 *Assume that we have a set of data following $N(\theta, \sigma^2)$ with unknown mean θ and standard deviation σ, and we are interested in testing $\theta = 0$ versus $\theta \neq 0$. Notice that the model assumption here is that the data follows a normal model with unknown mean and standard deviation, and we want to make a prediction on the asserted θ value.*

Under the above setting, the null space is $\theta \in \Theta_0 = \{\theta : \theta = 0\}$, and the alternative space is $\theta \in \Theta_1 = \{\theta : \theta \neq 0\}$.

We usually use the Student-t statistic as the test statistic

$$T_{n-1} = (\overline{X} - 0)/\frac{s}{\sqrt{n}},$$

where $\overline{X}$ is the sample mean and s is the sample standard deviation

$$S^2 = \frac{1}{n-1}\sum(X_i - \overline{X})^2.$$

To evaluate the performance of the test statistic, conventionally we use the following two criteria,

Definition 1.2 P(Type-I error): *The conditional probability of incorrectly rejecting the null hypothesis.*

Definition 1.3 P(Type-II error): *The conditional probability of incorrectly not-rejecting the null hypothesis.*

Usually, the probability of Type-II error is reported by its counterpart, the power of a test.

Definition 1.4 Power: *Power$= 1 - P(Type-II\ error)$, which is the conditional probability of correctly rejecting the null hypothesis. Namely the probability of making a correct decision by rejecting the false null hypothesis.*

Thus a way to evaluate a test statistic is to examine its probability of type-I error and the power of the test. We usually fix the probability of type-I error at level α, and use the concept of the power function to evaluate a test with a rejection region **R** for a given significance level. In the case where the alternative space contains more than one element, we use the concept of a power function to measure the probability of Type-II error when the population mean travels in the alternative space. When the null space contains more than one element, we use the minimal possible probability of Type-I error among probabilities associated with all parameters in the null space. This can be conveniently defined by the concept of a power function.

Definition 1.5 Power function: *The power function corresponding to a hypothesis test with rejection region **R** is the function defined on the domain of the parameter θ, which takes the form of $\beta(\theta) = P_\theta(X \in \mathbf{R})$, where θ is the parameter of primary interest in the hypothesis.*

With above definition, conventionally as defined in the literature, a test with power function $\beta(\theta)$ is a size α test, if

$$\sup_{\theta \in \Theta_0} \beta(\theta) = \alpha.$$

A test with power function $\beta(\theta)$ is a level α test, if $\sup_{\theta \in \Theta_0} \beta(\theta) \le \alpha$.

Obviously, the above analysis is grounded on the assumption of the underlying model. If the model can not be plausibly assumed, there is no test statistic. Under this scenario, if the sample size is large enough, the data-oriented approach can be applied. In what follows in this section, as an introduction to the evaluation criteria for data-oriented analytic approach, we shall discuss the evaluation criterion of a data-driven method in the next subsection.

1.4.2   Cost functions in neural networks

When the model of the underlying data can not be plausibly assumed, an alternative is to skip the model assumptions of the data and directly make inference without any assumption on the underlying data model. This data-oriented approach can be applied, especially when the data is large. For instance, one may partition the data into a training set (usually 75 percent of the data) and a testing data set (usually 25 percent of the data), and train the model use the training data, then evaluates the trained model with the testing data.

Similar to the preceding subsection using hypothesis testing as an illustrating example, in this subsection, we use neural network as an example to illustrate the evaluation of data-oriented inference. As suggested by its name, the fundamental principle of neural networks is inspired by the biology of the human brain. Neural networks constitute a substantial part of artificial intelligence. For instance, genetic algorithms are built upon evolution and particle swarm optimization is based on animal social behaviors. Generally speaking, a neural network is a directed graph, with neurons as nodes and weights as edges. Every neuron activates, or outputs, with a strength that is a function of the element-wise multiplication of the inputs with the edge weights.

Mathematically, neural network is a composite function of the input information, as defined in Goodfellow (2016)[51] and Lee (2017) [79]. In another word, define w_{ij} the weight value between nodes i and j, n_j the activation of node i, and N_j the list of node indexes that are connected to j, then node j will activate with strength

$$n_j = F(\sum_{i \in N_j} w_{ij} n_i),$$

where F is the activation function that allows the network to produce nonlinear behaviors.

We usually provide input into the neural network by activating a set of nodes with specific values and read output from any subset of nodes similarly. These networks are typically organized in layers of neurons, which indicate the depth of each node. In this way, layers are typically fully connected, meaning that all nodes in one layer are connected to all nodes in the next layer. This allows a computationally efficient model of weights as a matrix M, taking input vector V to output vector MV. For example, feed-forward networks are constructed in the following way. Its structure is often considered fixed and serves to provide a final classification. Key limitations of fully connected layers prevent them from being suitable for use as the sole structure of larger networks. Because of the fully connected nature of the layers, they require an immense amount of memory. Such a layer between two sets of just 10,000 nodes would require 100 million parameters, while modern networks often have a total of 10 million parameters. This extrema capacity, while being inefficient can also be problematic for training in general. There is no sense of locality in such layer, as every node is treated individually. This means that it is difficult and nearly impossible to train higher-level features that should be treated equally across all areas of the input (which is of particular interest to problems like image classification). However, even with these limitations, fully connected layers remain critical for the task they perform.

Besides the feed-forward network, the other key component of neural networks includes back propagation, which is an algorithm to let errors accumulated from the output layer of the network propagate backward through the network, training it in the process. As in the example, if the network's output is O, but the correct response would be C, we can calculate the error

$E = O - C$. From this, we need a cost function that determines how errors are judged, a typical example may be the L_2 loss

$$Cost(O - C) = \sum_{i=0}^{n} ||O_i - C_i||^2.$$

However, since we know that

$$O = F(\sum_{i=0}^{n} w_i a_i),$$

it is possible to figure out the influence each weight had on the error by taking the partial derivative of the cost function with respect to the weight.

Utilizing this partial derivative, each weight can be modified as a result of the preceding layer.

1.5   Learning process optimization

Following the introduction of the evaluation criteria for the two cultures of data analysis, we shall introduce general principles in the optimization process in terms of improving the inference performance, which, for instance, consists of the probability of type-I and type-II errors in hypothesis for model-based camp, or testing error rates in data-driven camp.

For the model-based culture camp, we will still use hypothesis testing as an introductory example to illustrate general strategies in controlling the Type-I error and optimizing the power, which is equivalent to lowering the probability of Type-II errors. Similar discussions can be applied to estimation, predictions, and classifications. For the data-oriented camp, we will use convolutions and deep-learning as introductory examples to discuss the optimization process. More details on optimizing strategies pertaining to specific topics will be addressed later in the corresponding chapters.

The process of statistical learning involves correct identification of appropriate data science camp to learn from the data with optimal learning outcomes according to the associated evaluation criteria.

1.5.1   Model-based camp

In this subsection, we use hypothesis testing as an example to illustrate optimization process for model-based culture camp. The optimization criteria discussed in the previous section serve as standards to seek (or approximate) the testing procedure with optimal performance. For example, when the underlying model is assumed for a hypothesis testing problem, the two evaluation criteria are the *Probability of Type-I error* and the *Probability of Type-II error*. A test procedure with "best" performance will be the one that reaches the highest permissible power with rejection region **R** at a pre-specified significance level α.

Under this setting, the issue of finding an optimal inference procedure becomes to finding the most powerful test for a given significance level. There are rich references at various levels in the literature in this regard, for instance Lehmann and Casella [80]; Casella and Berger [16], as well as Lehmann and Romano [81], to list just a few. We will briefly discuss basic results here to facilitate the understanding on the strategy in the construction of optimal learning procedures for model-based culture camp. More systematic details can be found in classical statistics literatures such as [16] or [81].

For a set of data (random sample) $X = (x_1, ..., x_n)^T$, assume that we are interested in determining whether the underlying model is $f(x|\theta_1)$ or $f(x|\theta_0)$. The optimal learning strategy toward a most powerful (MP) test for a given significance level α is the Neyman-Pearson fundamental lemma.

Theorem 1.2 Neyman-Pearson Lemma: *Assume that the underlying model (pdf or pmf) of a set of data X is $f(x|\theta) \in \{f(x|\theta_0), f(x|\theta_1)\}$. Namely there are only two candidate models, one is for the null hypothesis and the other for the alternative hypothesis. Consider testing $H_0 : \theta = \theta_0$ vs $H_1 : \theta = \theta_1$. The level α most powerful test is the one with the rejection region **R** satisfying*

$$X \in \mathbf{R} \text{ if } f(X|\theta_1) > kf(X|\theta_0);$$

$$X \in \mathbf{R}^c \text{ if } f(X|\theta_1) < kf(X|\theta_0),$$

for some constant $k \geq 0$ so that $P_{\theta_0}(X \in \mathbf{R}) = \alpha$.

In what follows, we shall use an example to illustrate the application of the Neyman-Pearson Lemma towards optimal learning procedures in model-based culture camp.

Example 1.5 *Let X be a random variable associated with diabetes symptoms. Assume that from historical data of patient records, the likelihood of each symptom under H_0 (blood glucose level ≤ 100) versus H_1 (blood glucose level > 100) for patients in a specific hospital are given in the following table.*

x	x_1	x_2	x_3	x_4	x_5	x_6	x_7	x_8
$f(x|H_0)$	.02	.02	.02	.01	.05	.01	0.41	0.46
$f(x|H_1)$	.22	.02	.12	.04	.18	.02	.06	0.34

Notation: x_1-numbness; x_2- weight loss; x_3-swollen gums; x_4- slow healing; x_5-increased appetite; x_6-blurred vision; x_7-energy loss; x_8-frequent thirst.

Based on the symptoms of the patient, we are interested in diagnosing whether a newly admitted patient has diabetes, with a restriction that the chance of incorrectly diagnosis a healthy patient as having diabetes is 5%.

Solution: Notice that in this case, the ratio of likelihood for each symptom $\lambda = \frac{f(x|H_1)}{f(x|H_0)}$ takes the following values

x	x_1	x_2	x_3	x_4	x_5	x_6	x_7	x_8
$f(x|H_0)$	0.02	0.02	0.02	0.01	0.05	0.01	0.41	0.46
$f(x|H_1)$	0.22	0.02	0.12	0.04	0.18	0.02	0.06	0.34
λ	11	1	6	4	3.6	2	0.146	0.739

Thus, on the basis of Neyman-Pearson lemma and according to the ranking of the likelihood ratio for each symptom, we arrange the symptoms by the likelihood of diabetes verse non-diabetes. To satisfy the evaluation criterion of controlling the rate of mis-diagnosis at 5% level, namely $P(X \in \mathbf{R}|H_0) = 0.05$, symptoms with the top 5% likelihood ratios are $\mathbf{R} = \{numbness, swollen\ gums, slow\ healing\}$.

The chance of correctly diagnosing a diabetes patient, which is the power of this MP test at size 0.05 reads

$$Power = P(X \in \mathbf{R}|H_1) = 0.22 + 0.12 + 0.04 = 38\%.$$

Certainly, when the blood glucose level test is available, the laboratory test result is more accurate in detecting diabetes as a follow-up diagnosis.

The above example makes inference on an unknown parameter for assumed models. The intuition is essentially the likelihood ratio test that optimizes the evaluation criterion by maximizing the power of the test while controlling the probability of incorrectly rejecting the null hypothesis. The rationale of the Neyman-Pearson lemma is grounded on the intuition that we reject the null hypothesis if the parameter in the alternative space is more likely to occur.

1.5.2   Data-Driven camp

A convolution neural network is the product of chaining together convolutions to perform efficient feature extraction with the standard feed-forward neural network structure. It is shown ([78]) that the same back-propagation methods used to train other networks could also be applied to convolution layers, allowing convolution neural networks to learn their own feature extractors.

This allows the convolution neural networks to determine what kinds of high-level feature extraction is necessary for the specific problem. More importantly, this allows for networks to automatically chain convolution layers, in which the initial information can pass through multiple layers of feature extraction, which are all automatically determined from the training data. Using a convolution kernel to pre-process the image proves to be critical to the performance of modern deep learning methods, as a small kernel can operate over a large image in parallel (see, for example, [79]).

In mathematics, a convolution is essentially an outcome of interpreting one function in the context of the other via the following formulation,

$$(f * g)(t) = \int_{-\infty}^{\infty} f(r)g(t - r)dr$$

From the perspective of modern deep learning, the primary use of the convolution technique is confined to the range of the convolution kernel $g(.)$ over $(0, s)$, as follows.

$$(f * g)(t) = \int_{0}^{s} f(r)g(t - r)dr.$$

For example, consider the basic edge-detecting matrix,

$$E = \begin{bmatrix} 0 & 1 & 0 \\ 1 & -4 & 1 \\ 0 & 1 & 0 \end{bmatrix}.$$

This convolution will perform the element-wise matrix multiplication of the kernel E with the immediate neighbors of each pixel, and then aggregate the elements by summation. That is, if the pixel values around a specific pixel e are

$$P_e = \begin{bmatrix} a & b & c \\ d & e & f \\ g & h & i \end{bmatrix},$$

the convolution at that pixel will be

$$
\begin{aligned}
P_e * E &= 0a + 1b + 0c + 1d - 4e + 1f + 0g + 1h + 0i \\
&= (b + d + f + h) - 4e.
\end{aligned}
$$

Accordingly, this creates a new matrix, with each element representing the convolution kernel applied at that point. As shown above, the convolution $P_e *$ E will have the strongest activation where there is a strong difference between the pixel e and its neighbors (b, d, f, h), thus performing a basic localized form of edge detection.

Summary
The impact of the books in statistical learning ([56], [72], [119], and [120]) and deep learning ([5], [51]) has penetrated into various applied fields including medical research ([1], [96], [97], [98], [111], and [116] among others). Methods in statistical learning essentially challenges the classical statistical theory and methodologies with seemingly discernible boundaries. In this chapter, we follow the idea of two cultures in statistical modeling by Breiman, Diaconis,

and Efron ([9], [10], [11], [13], and [44]) to analyze differences and intrinsic connections between the two culture camps in data science.

Based on the background training, knowledge, believe, and experience, it is debatable on the correct way of data analytics because such a uniformly correct decision does not exist. Instead of sailing in one direction on methodologies in data analytics, this chapter goes through the two directions from the evaluation criteria to optimization processes. We use a numerical example on insurance premium to elucidate that blindly performing either one approach in data science may result in misleading conclusions. The model-based inference camp demands plausibility in model assumptions behind a set of data (see, for example, [22], [25], [28], [40], [42], [43],[52], [53] and [91], among others). On the other hand, the data-driven inference camp necessitates large sample size ([44], [51],[54], [79], among others).

There is a dilemma in the selection of the two data inference culture camps. Seeking to thoroughly clarify the differences partially motivates the compilation of this volume. The choice of the analytic culture camp should be grounded on the feature information of the data. For instance, although big data or data with high dimension is often regarded by some as the motivation for data-driven technologies, the problem on high dimension is actually due to the result that the sample covariance matrix is not positive definite with probability one when $n < p$ (see, for instance, Xie and Chen (1988 [126], 1990 [127]). Large dimension by itself is not an issue. For regression analysis in the model-based camp, most of the theorems start with k dimensions where k is *any* positive integer.

Once the framework of analytics is settled, the key component is the selection of the evaluation criteria and optimization procedures. We use UMP (uniformly most powerful test) as an illustrating example for model-based culture camp, and MEPE (mean expected prediction error) for data-driven culture camp for the selection of suitable analytical approaches. The chapter concludes with a discussion on the principle of optimization strategies for the two analytical culture camps in data science. The rest of the chapters basically follow the theme and road-map in the setting of this chapter to delineate the two inference camps in data science.

2

Fundamental Instruments

This chapter discusses basic elements of statistical prediction and machine learning. The learning process usually consists of three fundamental parts: *identifying data, building or training models,* and *evaluating models.* To see why data identification is one of the critical steps in the learning process, consider the following simple example.

Example 2.1 *Assume that we have 15,000 breast cancer patients and 15,000 healthy participants in a case-control dataset. If we do not pay attention to the way in which the dataset was collected, it is easy to incorrectly claim that the disease rate is*

$$P(D) = 15000/(15000 + 15000) = 50\%.$$

In fact, some software even generates such a disease rate automatically. Notice that this is misleading because the numbers of cases and controls are prefixed before data collection in case-control data. The pre-determined 50% comes from the design stage in data collection. It has nothing to do with the disease risk $P(D)$, regardless of the sample size, learning methods, or testing methods. Correctly identifying the feature of the dataset helps us to select the appropriate approach, and avoid making commonly misleading errors in statistical learning.

Since the primary resource in the learning process is data, in this chapter we start with a discussion on the identification of different types of data. Certainly, in integrated part of data science is computation. However, a computer is, overall, just a machine that runs codes to carry out complicated computational tasks. If the learning method does not match well with the input data, the corresponding output mechanically produced by the machine could completely lead to the wrong direction. For instance, in the above breast cancer example, if a set of case-control data is mistreated as a set of cohort data, the learning result will be completely fallacious. The first issue in statistical learning is the understanding of the background information, features and characteristics of the input data so that the corresponding learning outcomes can be properly formulated.

Besides data identification and conventional learning methods such as regression, estimation and testing, other essential instruments in the implementation of machine learning include regression trees and classification trees. We shall also introduce the concept of decision trees as the second theme in this chapter to facilitate related discussions in follow-up chapters.

With a set of well-identified data, appropriately formulated prediction parameters (targets of learning outcomes), and implementation instruments such as computing algorithms in regression trees, the third theme in this chapter focuses on essential features of model evaluation, which includes sensitivity and specificity, ROC curves, cross-validation, and bootstrapping. ROC curves evaluate the plot of sensitivity and specificity for different threshold on classifications. Cross-validation is a critical approach that makes fine-tunning on models learned from training data, while bootstrapping is a data-driven approach that captures unknown features by re-sampling algorithms.

2.1 Data identification

The type of data dictates the learning tools and directions in prediction. We usually assume that the data constitute a random sample which fairly represents the population of interest. However, not all the data are random samples as seemingly assumed. Noticing that the type of data critically affects the learning method and the interpretation of the learning results. For instance, statisticians developed various analytic tools according to different sampling methods in experimental designs. On the other hand, real-life scenarios often do not permit the collection of data by random experiments, such as performing drug efficacy and toxicity experiments directly on human beings.

To see the point mentioned above, consider the analysis on the relationship between smoking and lung cancer. It is ethically inappropriate to conduct an experiment that exposes young children to cigarettes in order to test whether smoking or nicotine intake stimulates the development of lung cancer in children. In fact, most of our data (especially big data) are observational data. Different from experimental data, observational data mainly come in three different ways, *case-control data*, *cohort data*, and *cross-sectional data*, according to the approach in sample collection. In what follows in this section, we will use databases in medical studies to illustrate the importance on selecting appropriate learning approaches and prediction procedures according to features of the data.

2.1.1 Data types

Definition 2.1 Case-control data*: Case-control data are collected retrospectively according to the health status of patients. The case group contains retrospective features of diseased subjects, while the control group contains retrospective medical records of healthy subjects. The goal of a case-control study is usually to investigate potential risk factors associated with the disease.*

Case-control design is an effective way to collect data retrospectively to avoid making experiments on recruited subjects. It is used for the prediction on the odds ratio of a disease risk under two risk factors. It normally starts with a prefixed number of cases (disease population) and controls (non-disease population), and retrospectively traces down characteristics, features, and risk status of the patients to learn (or predict) the unknown risk of a disease. Case-control study often appear in clinical trials or medical investigations on risk factors associated with diseases such as stroke, breast cancers, or lung cancers, to list just a few. More information on this regard can be found in [1], [18], [111], and [116], among others. To control possible confounding effects, some clinical trials use matched case-control data where subjects are matched with demographic factors such as age, gender, or race.

Definition 2.2 Cohort data: *Cohort data are the data collected prospectively over a period of time. A cohort study usually recruits two groups of participants. One is exposed to the risk factor of interest, and another is not. The recruited participants are observed over time a period of time for records on disease development or occurrence of disease symptoms.*

Cohort study is more reliable in the causal relationship between the exposed and the diseased populations because it observes the development of the disease over time. However, it usually takes a long time and requires large sample sizes to control possible confounding factors and sampling bias. More examples of medical studies related to prospective studies can be found in, for example, [97], [98], and [99].

Definition 2.3 Cross-sectional data: *Different from case-control data (where the total numbers of cases and controls are fixed before sampling), or cohort data (where the numbers of exposure and non-exposure are fixed prior to the beginning of the study), a cross-sectional data only fixes the total number of subjects in the study in a single time point or a fixed location. Subjects involved in a cross-sectional data are assumed to form a random sample that fairly represents the surveyed population.*

For example, in a medical survey of 200 diabetes patients, the population is diabetes patients. If 30 out of 200 patients have hypertension in the survey outcome, the estimated prevalence for hypertension is 15%. If 40 out of 200 patients have smoking history, the estimated smoking rate is 20%. This is because in cross-sectional data, we do not pre-fixed the number of smoking or hypertension patients.

In terms of sampling cost, collecting cross-sectional data may be less expensive since it only involves surveys. On the other hand, since the number of diseased subjects and healthy subjects are unknown, the data may become useless when the number of cases (or controls) is too small to learn anything toward the feature of interest.

The method of data collection also dictates the corresponding analytical tools in statistical learning and prediction. As the primary source of information, data critically influence the learning outcome. To see this point, consider the analysis of the risk of a disease with a set of case-control data where the number of cases and the number of control are determined before the data collection stage. Regardless of the sample size and learning methods, the case-control dataset does not lead to any information for the prediction of the prevalence of the disease. Similarly, in a cohort study that contains prefixed number of patients exposed (and nonexposed) to a potential risk factor, it is methodologically fallacious to use such data to predict the prevalence of risk exposure, $P(E)$. This is because the ratio of risk exposure is given in cohort data before sampling, the same reasoning as to the prediction of disease rate $P(D)$ with case-control data. Identifying the dataset correctly helps us avoid making misleading conclusion in statistical learning.

Definition 2.4 Invariant measurement*: Let $m(\mathbf{X}|D)$ be a measurement based on data $\mathbf{X}$ obtained by the data collection method D. For two different data collection methods D_1 and D_2, if $m(\mathbf{X}|D_1) = m(\mathbf{X}|D_2)$, the data measurement $m(\mathbf{X})$ is called invariant for the two data collection methods D_1 and D_2.*

Example 2.2 *Consider the sample disease rate, $\hat{P}(case)$, where the measurement*

$$m(\mathbf{X}) = \frac{number\ of\ cases}{total\ sample\ size}.$$

Since

$$P(disease|case - control\ data) \neq P(disease|cohort\ data)$$

The prediction of disease rate, which is the sample proportion, $m(\mathbf{X})$, is not invariant between case-control data and cohort data.

Definition 2.5 Sample odds ratio*: Consider a data set in a 2×2 contingency table*

	Disease	Healthy	Total
Exposed	a	b	m_1
Nonexposed	c	d	m_2
Total	n_1	n_2	n

The sample odds ratio of disease for exposure patients is defined as,

$$\hat{OR} = \frac{ad}{bc}.$$

Interpretation of the sample odds ratio: If the sample odds ratio is around 1, exposing to the risk factor does not affect the odds of getting the disease; if the odds ratio is larger (less) than 1, exposing to the risk factor increases (decreases) the odds of getting the disease. Due to randomness behind the data, we usually claim significance of the odds ratio at 0.05 significance level when the 95% confidence interval of the odds ratio completely locates within the set $(-\infty, 0)$ or $(0, \infty)$.

The following theorem shows that the sample odds ratio is invariant between case-control data and cohort data. Thus, the sample disease prevalence rate depends on the data collection method, but the sample odds ratio is invariant. This means that as long as we use the sample odds ratio to measure the association between the disease and risk factors, the value of the sample odds ratio is invariant.

Theorem 2.1 *The sample odds ratio is invariant between case-control data and cohort data.*

Proof. Denote $P(D|E)$ the probability of disease in the exposure group, $P(D|E^c)$ the probability of disease in the control group, $P(E|D)$ the probability of exposure in the disease group, and $P(E|D^c)$ the probability of exposure in the non-disease group.

For case-control data, we have the sample odds of exposure in the case group,

$$\frac{\hat{P}(E|D)}{\hat{P}(E^c|D)} = \frac{\frac{a}{n_1}}{\frac{c}{n_1}} = \frac{a}{c}.$$

And the sample odds of exposure in the control group,

$$\frac{\hat{P}(E|H)}{\hat{P}(E^c|H)} = \frac{\frac{b}{n_2}}{\frac{d}{n_2}} = \frac{b}{d}.$$

Thus, the estimated odds ratio for the case-control data reads

$$\hat{OR}_{case-control} = \frac{a/c}{b/d} = \frac{ad}{bc} \tag{2.1}$$

For cohort data, we have the sample odds of exposure in the exposed group,

$$\frac{\hat{P}(D|E)}{\hat{P}(D^c|E)} = \frac{a}{m_1} \Big/ \frac{b}{m_1} = \frac{a}{b}.$$

And the sample odds of disease in the non-exposure group,

$$\frac{\hat{P}(D|E^c)}{\hat{P}(H|E^c)} = \frac{c}{m_2} \Big/ \frac{d}{m_2} = \frac{c}{d}.$$

Thus, the estimated odds ratio for the cohort data reads

$$\hat{OR}_{case-control} = \frac{a/b}{c/d} = \frac{ad}{bc} \tag{2.2}$$

Comparing the equations (2.1) and (2.2) gets the conclusion of the theorem.

Although the estimate of the disease prevalence is not invariant between cohort and case-control studies, Theorem 2.1 shows that the sample odds ratio is invariant between case-control study and cohort study. In fact, the population odds ratio is also invariant, by the use of conditional probability argument.

$$Population\ Odds - Ratio = \frac{\frac{P(D|E)}{1-P(D|E)}}{\frac{P(D|E^c)}{1-P(D|E^c)}} = \frac{\frac{P(E|D)}{1-P(E|D)}}{\frac{P(E|D^c)}{1-P(E|D^c)}}.$$

Thus, for the evaluation of disease risk related to an exposure, it is legitimate to use odds ratio for case-control and cohort study, instead of directly using the estimate of disease prevalence.

For cross-sectional data, the learning method on odds ratio is different from case-control data and cohort data since the marginal sums are random variables. We may use the conditional non-central hyper-geometric model to make inference on the odds ratio of disease between the exposed and unexposed populations.

There are also other types of data (such as time series data, survival data, longitudinal data, etc) that necessitate specific analytical/learning approaches for the characteristics and features of the data. To avoid misleading prediction outcomes in learning the model behind the data, it is critical to correctly identify the type of data and appropriately select the corresponding analytical approach.

2.1.2   Pooling data, Simpson's paradox, and solution

It is very common in data science to pool several sets of data together in the process of data analytics. One of the common mistakes, which is often overlooked in practice, is the impact of confounding factors that may alter the prediction outcome.

In statistical analysis, when dividing data from a population into subpopulations, this phenomenon is called the Simpson's paradox. However, similar effects also occur in data pooling. To further illustrate this point, we examine the following numerical example.

Example 2.3 *The following is a set of hypothetical data summarizing a survey (cross-sectional) data regarding residents' opinions on a new health policy. We have three features (variables) in the dataset.*

Resident community: Urban or rural;

Opinion: favoring or against;

Mental stress status: stressed or not stressed.

For the odds ratio measuring the relationship between being stressed and favoring the newly proposed healthy policy, we examine three datasets to see the change of odds ratios in the individual and collapsed datasets.

The odds ratio for the urban population reads,

$$OR_{urban} = \frac{48 \times 94}{12 \times 96} = 3.9167.$$

TABLE 2.1

Pooling data and Simpson's paradox

		favoring	not favoring
Urban	Not stressed	48	12
	Stressed	96	94
Rural	Not stressed	55	135
	Stressed	7	53

for the rural population,

$$OR_{rural} = \frac{55 \times 53}{135 \times 7} = 3.0847.$$

Both the odds ratios for the urban and rural populations are larger than 1, indicating that stress level influences the opinion on supporting the new health policy. However, when we collapse the data into one table, we have

	Favoring	Not favoring
Not stressed	103	147
Stressed	103	147

Thus, the odds ratio calculated from the collapsed data is 1, which implies that being stressed does not have any impact on the opinion of the new health policy.

Now, we have two contradicting statements with the odds ratio for the impact of being stressed on political opinions. One claims that being mentally stressful has impact on residential opinion for the public health policy, and the other does not.

In fact, it is implausible to pool the data sets together in Table 2.1 because one is for urban population while the other is for the rural population. The confounding effect of community locations alters the relationship between stressful population and opinion on public health policy. In this situation, it is more appropriate to use the *Cochran-Mental-Haemsel* approach for the odds ratio when combining data information from two different sources with confounding effects.

$$OR_{CMH} = \{\sum_i \frac{a_i d_i}{N_i}\}/\{\sum_j \frac{b_j c_j}{N_j}\},$$

where the dataset has the following setting for the k data sources with $i\,j = 1, ..., k$.

With the adjustment of confounding factors in each data stratum (data source in Table 2.1), the adjusted odds ratio reads

$$OR_{CMH} = \frac{48 \times 94}{250} + \frac{55 \times 52}{250} / \frac{12 \times 96}{250} + \frac{135 \times 7}{250} = \frac{7427}{2097} = 3.5417,$$

which is consistent with the conclusion on the impact of odds ratio in each

TABLE 2.2
OR and CMH for Simpson's paradox

	Case	Control	Total
Exposure	a_i	b_i	n_{i1}
Nonexposure	c_i	d_i	n_{2i}
Total	m_{1i}	m_{2i}	N_i

data stratum that people with stress tend to against the new health policy while those without stress are in favor of the new policy.

Certainly, the hypothetical dataset is constructed in the way to amplify the confounding effect. However, it points out the fact that pooling datasets (with the frame of data similar to Table 2.2) without carefully considering potential confounding factors may completely alter the learning outcome, and consequently result in misleading conclusions. Confounding effects in Simpson's paradox necessitates adjusting methods such as the CMH weighting approach.

2.2 Basic concepts of trees

After identifying the nature of the data, the next step is to find an appropriate learning method to bridge the data and the unknown features for prediction. This process involves functions of the data (statistics) in conjunction with measurements of data variation (probability). Since probability and statistics are very well developed and documented in the literature. In this section, we focus on the discussion of basic concepts and fundamentals of a relatively new topic, the decision tree.

Different from conventional methods, the use of decision trees partly marks a distinct feature of machine learning. It usually includes regression trees and classification trees. When we have a set of explanatory variables to predict a dependent variable, the conventional approach in statistics, for example, is the linear regression

$$Y = a + b_1 X_1 + ... + b_k X_k + \epsilon,$$

where ϵ is a random variable corresponding to the distribution of Y. In this setting, the variables X_1, ..., X_k are assumed to be equally affecting the responsible variable Y.

$$E(Y|\mathbf{x}) = a + b_1 X_1 + ... + b_k X_k,$$

where the vector $\mathbf{x} = (X_1, ..., X_k)^t$. However, in practice, there is a high possibility that one of the explanatory variables is more prominent in determining the value of Y, as shown in the following example.

Example 2.4 *Consider a scenario in marketing analysis where Y is the sale volume, X_1 is the advertising input, and X_2 is the selling price. In a regression model, we are unable to claim whether advertising input should be considered before the selling price, or vice versa. This problem can be resolved by a regression tree.*

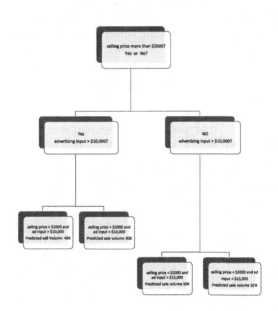

FIGURE 2.1
Regression tree of sale volume on price and advertising cost

As shown in Figure 2.1, the first split of the regression tree is on whether the selling price is more than \$2000, we essentially consider the selling price first. Products with selling price more than \$2000 will be considered in one category (branch), while the advertising input will be considered as the second criterion (sub-branch). If the first split is on advertising cost, we will correspondingly consider the advertising cost first.

Interpretation: For the branch where the selling prices of the product are more than \$2000, the advertising input will be considered. Assume that the split point for the advertising input is \$10,000. We essentially consider two subbranches for products in the first branch: one subbranch has selling price more than \$2000 and advertising input more than \$10,000; while another subbranch is for products with selling price more than \$2000 and advertising input no more than \$10,000. Since products in each subbranch share the same impact from the two factors (selling price and advertising input), the sale volumes for products in each subbranch are averaged for the predicted sale volume in the terminal node, $(40K, 30K)$.

For products in the branch where selling prices are no more than \$2000, similar consideration leads to the following two sub-branches. The first sub-branch consists of products with selling prices no more than \$2000 and advertising input more than \$15,000, while the second subbranch consists of products with selling prices no more than \$2000 and advertising input no more than \$15,000, correspondingly. And the sale volumes of products in each subbranch are averaged up for the predicted sale volumes, 50 K and 10 K, respectively.

In the description of the construction of a regression tree in Figure 2.1, as mentioned before, one of the distinct features is the order of the explanatory variables being considered in the prediction process. Another key issue is the determination of the split point for the explanatory variable under consideration. It is related at this point to introduce two basic definitions in the construction of a regression tree.

Definition 2.6 Feature Space*: The set of all possible input combinations of explanatory variables in statistical prediction.*

For example, when we consider X_1 selling price and X_2 advertising input of the company, the feature space is

$$S = \{(X_1, X_2) \in R^+ \times R^+\}.$$

The concept of feature space is closely related to the concept of feature space partition in the theory of decision trees.

Definition 2.7 Partition of the feature space*: Let Θ be the feature space of a set of predictors. Denote $\{\Theta_1, ..., \Theta_k\}$ the set of mutually exclusive subsets of the feature space Θ, such that*

$$\Theta_i \bigcap \Theta_j = \emptyset, \ i \neq j \quad and \quad \bigcup_{i=1}^{k} \Theta_i = \Theta.$$

The set $\{\Theta_1, ..., \Theta_k\}$ is called a partition of the feature space Θ.

For instance, for any two positive values $s_1 > 0$ and $s_2 > 0$, let

$$R_{11} = \{X_1 \leq s_1, X_2 \leq s_2\} \quad R_{12} = \{X_1 \leq s_1, X_2 > s_2\}$$

$$R_{21} = \{X_1 > s_1, X_2 \leq s_2\} \quad R_{22} = \{X_1 > s_1, X_2 > s_2\},$$

The set $\{R_{11}, R_{12}, R_{21}, R_{22}\}$ is a partition of the feature space $[0, \infty) \times [0, \infty)$.

With a partition of the feature space under the assumption that subjects in the same set of a partition share the same expected response, we have

$$y_t = \sum_{x_t \in R_{ij}} c_{ij} I_{R_{ij}}(x_t),$$ (2.3)

where the indicator function

$$I_{R_{ij}}(x_t) = \begin{cases} 1 & \text{if } x_t \in R_{ij} \\ 0 & \text{otherwise.} \end{cases}$$

In this setting, the estimate of c_{ij} (which is the same as the expected response within the partition R_{ij}) with minimum mean squared prediction error reads

$$c_{ij} = \hat{y}_{ij}^* = \frac{1}{n_{ij}} \sum_{t:x_t \in R_{ij}} y_t,$$

where n_{ij} denotes the sample size in the set R_{ij}.

The following example explains the notation of the indicator function in equation (2.3).

Example 2.5 *Consider a set of data where the sample mean responses within each set of the partition are 15, 20, 30, and 40. Similar to the tree demonstrated in Figure2.1, the tree predicted model (2.3) becomes*

$$\hat{y}(x_t) = 15 I_{R_{11}}(x_t) + 20 I_{R_{12}}(x_t) + 30 I_{R_{21}}(x_t) + 40 I_{R_{22}}(x_t).$$

Depending on the set R_{ij} that the value x_t falls into, if $x_t \in R_{11}$, we have $I_{R_{11}}(x) = 1$. Since $\{R_{11}, R_{12}, R_{21}, R_{22}\}$ forms a partition of the feature space, the observation x_t now does not belong to any one of the sets R_{12}, R_{21}, or R_{22}, we have $x \notin R_{ij}$ when $i \neq 1$ or $j \neq 1$. Thus

$$I_{R_{12}}(x_t) = I_{R_{21}}(x_t) = I_{R_{22}}(x_t) = 0,$$

and equation (2.3) leads to

$$\hat{y}(x_t) = 15.$$

Example 2.6 *The process of constructing a regression tree is equivalent to an optimization process that minimizes the prediction error (E) under all possible partitions of the feature space,*

$$E = \sum_{i:(x_{1i},x_{2i}) \in R_{11}} (y_i - \hat{y}_1^*)^2 + \sum_{i:(x_{1i},x_{2i}) \in R_{12}} (y_i - \hat{y}_2^*)^2$$

$$+ \sum_{i:(x_{1i},x_{2i}) \in R_{21}} (y_i - \hat{y}_3^*)^2 + \sum_{i:(x_{1i},x_{2i}) \in R_{22}} (y_i - \hat{y}_4^*)^2$$

$$= \sum_{uv} \sum_{i:(x_{1i},x_{2i}) \in R_{uv}} (y_i - \hat{y}_{(uv)}^*)^2$$ (2.4)

*where R_{uv} is a set of the product in the subbranch (uv), $\hat{y}_i^*$, $i = 1, 2, 3, 4$
is the average sale volume for products in the subbranch R_{uv}, for $(uv) \in
\{(11), (12), (21), (22)\}$, respectively.*

*In Figure 2.1, the set $\{R_{11}, R_{12}, R_{21}, R_{22}\}$ serves as a partition of the
feature space $[0, \infty) \times [0, \infty)$.*

Different from the linear regression method where the random error ϵ follows a statistical model, tree regressions do not need any model assumption.
On the other hand, the difficulty on the implementation of regression tree
switches to the selection of the partition of the feature space, that minimizes
the mean squared prediction errors. For instance, in the selling price and advertising input example, the construction of a regression tree depends on the
selection of values s and t, as well as the order of the two explanatory variables,
X_1 and X_2. When the number of features (predictors) increases, the corresponding volume of computation will increase dramatically. This necessitates
the use of computing software for the construction of regression trees.

As a fundamental introduction to the basic concept of statistical learning,
this section discusses the concept and interpretation of the decision tree. More
details on this topic, especially the uniformly minimum variance unbiased
estimator (UMVUE) for the homogeneity index in each terminal node, will be
delineated in Chapter 9.

2.3   Sensitivity, specificity, and ROC curves

Given a set of training data, we can train a binary classifier (a decision rule)
for the diagnosis of disease or non-disease using features of the patients (a
set of explanatory variables). It is related at this point to mention four possible outcomes corresponding to a decision rule for binary classifications with
a diagnostic threshold. For each decision, if the prediction is negative, in the
case where the real feature is negative, there is no error in the prediction, the
outcome is *true negative*. On the other hand, if the real feature is positive,
there is an error of falsely claiming negative (*false negative*). Similarly, if the
prediction is positive, in the case where the real outcome is actually negative,
there is a misclassification error on incorrectly claiming positive (*false positive*). If the real feature is indeed positive, the prediction makes a true positive
assertion.

The four possible outcomes in binary classification can be summarized
in Table 2.3. For example, in the process of classifying a disease, there is a
possibility of misclassifying a diseased patient as healthy (false negative), or
misclassifying a healthy patient as having a disease (false positive). These
two errors are conceptually similar to the type-I errors and type-II errors in
hypothesis testing. Along the same line on the evaluation of discriminating

TABLE 2.3

Four possible outcomes in a binary classification

Feature	predicted negative	predicted positive	Total
Real negative	true negative	false positive	$n_{real\ negative}$
Real positive	false negative	true positive	$n_{real\ positive}$
Total	$m_{predicted\ negative}$	$m_{predicted\ positive}$	N

ability for a diagnostic test, two concepts frequently used in the literature are sensitivity and specificity.

Definition 2.8 Sensitivity*: Let T be a binary classifier trained with the training data. The probability that T correctly diagnoses a disease patient as having the disease is defined as the sensitivity of the classifier T. If larger value of T indicates higher likelihood of the disease, for a threshold value c,*

$$Sensitivity = P(T\ diagnoses\ a\ case|case) = P(T > c|case).$$

Clearly, sensitivity is the probability of correct diagnosis of a disease condition on the diseased population. It is essentially a conditional probability depending on a given threshold c.

Definition 2.9 Specificity*: Let T be a binary classifier trained with the training data. The specificity of the classifier T is defined as the probability that T correctly claims a healthy patient as healthy. For a threshold value d,*

$$Specificity = P(T\ claims\ a\ healthy\ outcome|healthy)$$
$$= P(T < d|healthy).$$

Clearly, the specificity is a probability of correct diagnosis conditioned on the healthy population. As the value d changes, the associated conditional probability changes.

For instance, when we use the logistic regression model to seek the odds of getting infected with a disease, given a set of features $\mathbf{x}$, assume

$$log(\frac{p}{1-p}) = \alpha + \beta^t\mathbf{x},$$

where the disease rate $p = P(Y = 1|\mathbf{x})$, and α and β are model parameters.

After training the model, we obtain the estimated values of α and β, denoted as $\hat{\alpha}$ and $\hat{\beta}$, and

$$\hat{P}(Y = 1|\mathbf{x}) = \frac{e^{\hat{\alpha}+\hat{\beta}^t\mathbf{x}}}{1 + e^{\hat{\alpha}+\hat{\beta}^t\mathbf{x}}} \tag{2.5}$$

The above equation predicts the chance of getting a disease for specific features of each patient, **x**. Thus, for each value c, based on the trained classifier, T, we can compute the estimated sensitivity by dividing the number of correct diagnosis with the total number of cases in the sample. Similarly, estimated specificity can be obtained for each value of c. Thus, in a trained model (binary classifier), a pair of values (sensitivity, specificity) can be computed for each diagnostic threshold c.

Considering all possible values of permissible diagnostic threshold c, $c \in A$, gets a set of pairs

$$\{(sensitivity_c, specificity_c), c \in A\}.$$

Plotting this set of data using the pairs $(sensitivity(c), 1 - specificity(c))$ for all $c \in A$ yields a curve, which is the ROC curve.

Definition 2.10 ROC curve: *The plot of (sensitivity, 1-specificity) across all permissible values of the diagnostic threshold c is called the receiver operating characteristic curve, or the ROC curve.*

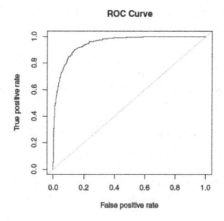

FIGURE 2.2
Example of a sample ROC curve

The ROC curve is basically the graphic plot of two parameters: sensitivity versus (1-specificity). When the value c travels in its domain determined by the diagnostic threshold. It evaluates the discriminating ability of the binary classifier. When a set of testing data is available, we may plug the data into the estimated model (2.5) to obtain a set of sample sensitivities and sample specificities associated with different cut-off (threshold) values of the classifier. The plot of the sample (estimated) sensitivity versus false positive rates (1-sample specificity) forms an estimated ROC curve, as shown in Figure 2.2.

2.4 Cross-Validation

After data identification, statistical learning of an unknown underlying model usually involves three portions, model training, model validation, and model testing. Correspondingly, the dataset is ideally split into three mutual exclusive subsets, a training set, a validation set, and a test set. The three portions are briefly described below.

The model training portion fits a candidate model with training data to estimate the model parameters. The selection of the candidate model usually is based on data analytic knowledge, for instance, we assume that the mean response is a linear function of the predictors. In this step, assumptions on the data and background information about the data play a critical role in the selection of the appropriate model. In machine learning processes, usually the training dataset fits a specific candidate model. However, there may be more than one candidate model, such as different orders in polynomial regressions, where we may fit a linear function or a quadratic function as candidate models. In polynomial regression, the degree of the polynomial model is usually validated with validation data.

The model validation part involves parameter estimation on the basis of an accuracy measurement in conjunction with the validation data, such as the selection of model coefficients corresponding to the smallest MSPE (mean squared prediction error). As for artificial neural networks, the number of hidden units in each layer in artificial neural networks is a hyper-parameter to be determined in the validation data. In general, the fine-tune of the trained model necessitates the validation process.

The model testing portion usually includes the evaluation of the final model with the testing data that was set aside to independently access the performance of the final model. It is inappropriate to estimate the predictive model and calculating the validation criteria to justify the final model because, overall, the validation data is just one random sample representing the population. Especially when it comes to prediction, the first step is to build or estimate a model which is then used to predict the unknown response.

One of the critical steps in the prediction process is to build a model that fits well with the data. Usually we spend 50% of the data training the model, 25% of the data on validation and 25% of the data on testing. However, when the sample size is not large enough for splitting into the training set and validation set, an efficient way is to cooperate the training and validating parts with the training data by the method of cross-validation.

Generally, cross-validation is a data implementation process that uses numerical computation to replace thorny theoretical analysis. It evaluates the trained model multiple rounds by different partitions of the training data, then takes the average of the corresponding evaluated model accuracy (such as the MSE) to give an estimate of the model's predictive performance. The cross-validation method usually provides a more accurate evaluation on fitting the model to the data.

Definition 2.11 Cross-validation Procedure: *Cross-validation divides the training data into k-equal size folds, trains the model using $k - 1$ folds of the data, and evaluates the trained model with the remaining fold of the data for the model accuracy.*

The validation process uses k different folds of the data, which consequently generates k accuracy measurements. Taking the average of the k accuracy measurements leads to the overall accuracy level of the model.

In practice, we usually use LOOCV (leave-one-out cross-validation), 5-fold CV, and 10-fold CV.

2.4.1   LOOCV and Jackknife

The LOOCV is actually n-fold cross-validation when the training data contains n observations. For each data point in LOOCV, the $n - 1$ observations are used to trained the model, and the remaining one data point is used to calculate the model accuracy. The average of n model accuracy measurements is then used as the index for the LOOCV model accuracy.

For example, when we use MSE as a measurement of model accuracy between the trained model and the data, the cross-validation accuracy reads,

$$CVA_{(n)} = \frac{1}{n} \sum_{i=1}^{n} MSE_i.$$

Consider the overall model MSE,

$$MSE = \frac{1}{n} \sum_{i=1}^{n} (\hat{y}_i - y_i)^2.$$

In LOOCV, the cross-validation model accuracy becomes

$$CVA(n) = \frac{1}{n} \sum_{i=1}^{n} (\hat{y^*}_i - y_i)^2,$$

where $\hat{y^*}_i$ is the predicted value of y_i with predictor x_i, corresponding to the model trained by the data $(x_1, y_1), ..., (x_{i-1}, y_{i-1}), (x_{i+1}, y_{i+1}), ...(x_n, y_n)$.

It should be noted that the drop-one observation feature of LOOCV shares the same sub-sampling principle as the jackknife replication procedure, where for a random sample $x_1, ..., x_n$, define an estimator based on a drop-one observation sample,

$$\hat{\theta}_{(i)} = f_{n-1}(x_1, ..., x_{i-1}, x_{i+1}, ..., x_n),$$

with $i = 1, ..., n$. And the jackknife estimate as the average of the n drop-one sample estimates,

$$\hat{\theta}_{Jack} = \frac{1}{n} \sum_{i=1}^{n} \hat{\theta}_{(i)}.$$

Example 2.7 *We shall use a toy dataset, $\{1, 5, 9\}$, to illustrate the cross-validation accuracy (CVA) for the prediction of the population mean using LOOCV, along with the jackknife estimation.*

Case-1 Since the overall sample mean is 5, we have the mean squared error without any data manipulation,

$$MSE = \frac{1}{3}[(5-1)^2 + (5-5)^2 + (9-5)^2] = \frac{32}{3}.$$

Case-2 Since there are three observations in the dataset, we have the three drop-one sample for the population mean,

$$\hat{\theta}_{(1)} = \frac{1}{2}(5+9) = 7$$

$$\hat{\theta}_{(2)} = \frac{1}{2}(1+9) = 5$$

$$\hat{\theta}_{(3)} = \frac{1}{2}(1+5) = 3,$$

which are fixed for $\hat{y}^*_1$, $\hat{y}^*_2$, and $\hat{y}^*_3$, respectively. The cross-validation accuracy,

$$CVA(3) = \frac{1}{3}[(7-1)^2 + (5-5)^2 + (3-9)^2] = \frac{72}{3} = 24.$$

Case-3 The jackknife estimator of the population mean,

$$\hat{\theta}_{(Jack)} = \frac{1}{3}(7+5+3) = 5,$$

which shares the same accuracy as case [1] above.

Theorem 2.2 *The jackknife procedure does not change the estimation on the population mean, but it reduces the variation of the data around the sample mean by $\frac{1}{n-1}$.*

Proof. Notice that

$$\hat{\theta}_{Jack} = \frac{1}{n} \sum_{i=1}^{n} \hat{\theta}_{(i)} = \frac{1}{n} \sum_{i=1}^{n} \frac{n\bar{x} - x_i}{n-1} = \frac{n-1}{n-1} \bar{x} = \bar{x}.$$

$$|\hat{\theta}_{(i)} - \hat{\theta}_{Jack}| = |\frac{n\bar{x} - x_i}{n-1} - \frac{1}{n}\sum_{i=1}^{n} x_i|$$

$$= |\frac{1}{n-1}(n\bar{x} - x_i - (n-1)\bar{x})|$$

$$= |\frac{1}{n-1}(\bar{x} - x_i)|$$

This completes the proof of theorem 2.2.

Theorem 2.3 *For any asymptotic unbiased estimator, the jackknife version of the estimator improves the convergence rate from $O(\frac{1}{n})$ to $O(\frac{1}{n^2})$.*

Proof: To see this, notice that if $T(X)$ is an asymptotic unbiased estimator of a function of the parameter $\tau(\theta)$ with convergence rate $O(\frac{1}{n})$, we have

$$E(T(X)) = \tau(\theta) + O(\frac{1}{n}).$$

Define $X_{(-i)}$ the jackknife duplicate (the sub-sample of $(n-1)$ observations excluding x_i). The jackknife version of $T(X)$ is

$$T_{Jack}(X) = nT(X) - \frac{n-1}{n}\sum_{i=1}^{n} T(X_{(-i)}),$$

the asymptotic performance of the estimator T_{Jack} reads

$$E(T_{Jack}(X)) = nE[T(X)] - \frac{n-1}{n}\sum_{i=1}^{n} E[T(X_{(-i)})]$$

$$= n[\tau(\theta) + O(\frac{1}{n})] - \frac{n-1}{n}\sum_{i=1}^{n}[\tau(\theta) + O(\frac{1}{n-1})]$$

$$= (n - n + 1)\tau(\theta) + nO(\frac{1}{n}) - (n-1)O(\frac{1}{n-1})$$

Now, notice that

$$nO(\frac{1}{n}) = a_1 + \frac{a_2}{n} + ...$$

and

$$(n-1)O(\frac{1}{n-1}) = a_1 + \frac{a_2}{n-1} + ...,$$

we have

$$nO(\frac{1}{n}) - (n-1)O(\frac{1}{n-1}) = a_1 + \frac{a_2}{n} + \dots - a_1 - \frac{a_2}{n-1} - \dots$$
$$= \frac{a_2}{n(n-1)}$$
$$= O(\frac{1}{n^2})$$

This completes the proof of theorem 2.3.

2.4.2 LOOCV for linear regressions

Since LOOCV is often used in linear regression, we discuss a theoretical simplification and example on LOOCV in multiple linear regression in this subsection. Consider the cross-validation of a linear model with a predictor row vector $\mathbf{x}'_i$ and a response y_i, $i = 1, ..., n$,

$$E(Y|\mathbf{x}_i) = \mathbf{x}_i \beta,$$

where β is the column of parameter vectors. Denote the response vector $\mathbf{y} = (y_1, ..., y_n)$, and the corresponding predictor matrix $\mathbf{X}$, we have the model

$$\mathbf{y} = \mathbf{X}\beta + \epsilon.$$

In general, when the sample size n is large, the regular LOOCV method is computationally expensive. However for the linear model, the cross-validation accuracy (CVA) with LOOCV can be simplified as in the following theorem.

Theorem 2.4 *When LOOCV is applied to linear regressions, the cross-validation accuracy (CVA) can be expressed explicitly as,*

$$CVA(n) = \frac{1}{n} \sum_{i=1}^{n} (\frac{y_i - \hat{y}_i}{1 - h_i})^2, \tag{2.6}$$

where $\hat{y}_i = \mathbf{x}'_i \hat{\beta}$, the estimated response for the ith observation, and h_i the leverage of the ith observation

$$h_i = \mathbf{x}'_i (\mathbf{X}'\mathbf{X})^{-1} \mathbf{x}_i.$$

Proof: Denote $\mathbf{X}_{(-i)}$ and $y_{(-i)}$, respectively, the data without the ith observation y_i and $\mathbf{x}_i$, $\hat{\beta}_{(-i)}$ is the vector of the least square estimate using $\mathbf{X}_{(-i)}$ and $y_{(-i)}$, and $\hat{y}^*_i$ is the estimated response using LOOCV with the ith observation dropped.

Note that under this setting,

$$CVA(n) = \frac{1}{n} \sum_{i=1}^{n} (y_i - \hat{y}^*_i)^2, \quad \hat{y}^*_i = \mathbf{x}'_i \hat{\beta}_{(-i)}.$$

Recall the following relationship in matrix algebra,

$$\mathbf{X}'\mathbf{X} = \mathbf{X}'_{(-i)}\mathbf{X}_{(-i)} + \mathbf{x}_i\mathbf{x}'_i \tag{2.7}$$

Under the assumption that $n - 1 > p$, see Xie and Chen (1988 [126]), the sample covariance matrix is positive definite with probability 1, both $(\mathbf{X}'\mathbf{X})^{-1}$ and $(\mathbf{X}'_{(-i)}\mathbf{X}_{(-i)})^{-1}$ exist. Multiplying $(\mathbf{X}'\mathbf{X})^{-1}$ in both sides of (2.7) yields,

$$\mathbf{I}_p = [\mathbf{X}'_{(-i)}\mathbf{X}_{(-i)}](\mathbf{X}'\mathbf{X})^{-1} + \mathbf{x}_i\mathbf{x}'_i(\mathbf{X}'\mathbf{X})^{-1}. \tag{2.8}$$

Notice that we also have

$$\mathbf{I}_p = (\mathbf{X}'_{(-i)}\mathbf{X}_{(-i)})(\mathbf{X}'_{(-i)}\mathbf{X}_{(-i)})^{-1},$$

which, in conjunction with (2.8), leads to

$$[\mathbf{X}'_{(-i)}\mathbf{X}_{(-i)}][\mathbf{X}'_{(-i)}\mathbf{X}_{(-i)}]^{-1} = (\mathbf{X}'_{(-i)}\mathbf{X}_{(-i)})(\mathbf{X}'\mathbf{X})^{-1} + \mathbf{x}_i\mathbf{x}'_i(\mathbf{X}'\mathbf{X})^{-1}. \tag{2.9}$$

Multiplying $[\mathbf{X}'_{(-i)}\mathbf{X}_{(-i)}]^{-1}$ from the left-hand side of (2.9) gets

$$[\mathbf{X}'_{(-i)}\mathbf{X}_{(-i)}]^{-1} = (\mathbf{X}'\mathbf{X})^{-1} + (\mathbf{X}'_{(-i)}\mathbf{X}_{(-i)})^{-1}\mathbf{x}_i\mathbf{x}'_i(\mathbf{X}'\mathbf{X})^{-1}. \tag{2.10}$$

Now, multiplying $\mathbf{x}_i$ from the right-hand side of (2.10) results in

$$[\mathbf{X}'_{(-i)}\mathbf{X}_{(-i)}]^{-1}\mathbf{x}_i = (\mathbf{X}'\mathbf{X})^{-1}\mathbf{x}_i + (\mathbf{X}'_{(-i)}\mathbf{X}_{(-i)})^{-1}\mathbf{x}_i\mathbf{x}'_i(\mathbf{X}'\mathbf{X})^{-1}\mathbf{x}_i, \tag{2.11}$$

which becomes

$$(\mathbf{X}'\mathbf{X})^{-1}\mathbf{x}_i = (1 - h_i)[\mathbf{X}'_{(-i)}\mathbf{X}_{(-i)}]^{-1}\mathbf{x}_i, \tag{2.12}$$

since the leverage of the ith observation is defined as

$$h_i = \mathbf{x}'_i(\mathbf{X}'\mathbf{X})^{-1}\mathbf{x}_i.$$

By (2.12), we have

$$(\mathbf{X}'_{(-i)}\mathbf{X}_{(-i)})^{-1}\mathbf{x}_i = \frac{1}{1 - h_i}(\mathbf{X}'\mathbf{X})^{-1}\mathbf{x}_i \tag{2.13}$$

Now, by the LSE estimates of the regression coefficients corresponding to $\mathbf{X}$ and $\mathbf{X}_{(-i)}$,

$$\mathbf{X}'\mathbf{X}\hat{\beta} = \mathbf{X}'\mathbf{y}$$

$$(\mathbf{X}'_{(-i)}\mathbf{X}_{(-i)})\hat{\beta}_{(-i)} = \mathbf{X}'_{(-i)}\mathbf{y}_{(-i)}$$

Consider (2.7), and $\mathbf{X}'\mathbf{y} = \mathbf{X}'_{(-i)}\mathbf{y} + \mathbf{x}_i y_i$, we have

$$(\mathbf{X}'_{(-i)}\mathbf{X}_{(-i)} + \mathbf{x}_i\mathbf{x}'_i)\hat{\beta} = \mathbf{X}'_{(-i)}\mathbf{y}_{(-i)} + \mathbf{x}'_i y_i,$$

where y_i is the response corresponding to the ith observation. Now, applying $(\mathbf{X}'_{(-i)}\mathbf{X}_{(-i)})^{-1}$ from left in both sides of the equality gets

$$\{\mathbf{I}_k + (\mathbf{X}'_{(-i)}\mathbf{X}_{(-i)})^{-1}\mathbf{x}_i\mathbf{x}'_i\}\hat{\beta} = \hat{\beta}_{(-i)} + (\mathbf{X}'_{(-i)}\mathbf{X}_{(-i)})^{-1}\mathbf{x}_i(\mathbf{x}'_i\hat{\beta} + \hat{e}_i),$$

where $e_i = y_i - \hat{y}_i$, the residual of the ith observation with β estimated by the complete data. Thus

$$\hat{\beta} = \hat{\beta}_{(-i)} + (\mathbf{X}'_{(-i)}\mathbf{X}_{(-i)})^{-1}\mathbf{x}_i\hat{e}_i.$$

By (2.13), the above equation can be simplified as

$$\hat{\beta} = \hat{\beta}_{(-i)} + (\mathbf{X}'\mathbf{X})^{-1}\mathbf{x}_i\frac{\hat{e}_i}{1 - h_i}.$$

and

$$\mathbf{x}'_i\hat{\beta} = \mathbf{x}'_i\hat{\beta}_{(-i)} + \mathbf{x}'_i(\mathbf{X}'\mathbf{X})^{-1}\mathbf{x}_i\frac{\hat{e}_i}{1 - h_i}. \tag{2.14}$$

Let $\hat{d}_{(i)}$ be the residual associated with the "leaving the i-th column's observation" data, we have

$$\hat{d}_{(i)} = \mathbf{y}_i - \mathbf{x}'_i\hat{\beta}_{(-i)},$$

and by (2.14),

$$\begin{aligned}
\hat{e}_{(i)} &= y_i - \mathbf{x}'_i\hat{\beta}_{(-i)} \\
&= y_i - \mathbf{x}'_i\hat{\beta} + h_i\frac{\hat{e}_i}{1 - h_i} \\
&= \hat{e}_i + (\frac{\hat{e}_i}{1 - h_i})h_i \\
&= \frac{\hat{e}_i}{1 - h_i}
\end{aligned}$$

which, in conjunction with

$$CVA(n) = \frac{1}{n}\sum_{i=1}^{n}\hat{d}^2_{(i)} = \frac{1}{n}\sum_{i=1}^{n}[\frac{y_i - \hat{y}_i}{1 - h_i}]^2.$$

This concludes the proof for (2.6).

In some occasions (such as SAS outputs), the statistic $CVA(n)$ is alternatively denoted as PRESS (prediction residual error sum of squares),

$$PRESS = \sum_{i=1}^{n}(y_i - \hat{y}^*_i)^2 = \sum_{i=1}^{n}[\frac{y_i - \hat{y}_i}{1 - h_i}]^2,$$

where $\hat{y}^*_i$ is the estimated response using LOOCV with the ith observation dropped.

```
data Fresh;
set WORK.'Fresh_multiple regression data'n;
x1=log('sale price'n);
x2=log('competitor price'n);
x3=log('advertising cost'n);
x4='sale price'n;
x5='competitor price'n;
x6='advertising cost'n;
run;
proc glmselect;
  model 'market demand'n= x1-
x6/selection=forward(stop=CV) details=steps
cvMethod=split(117);
  run;
```

Data Set	WORK.FRESH
Dependent Variable	market demand
Selection Method	Forward
Select Criterion	SBC
Stop Criterion	Cross Validation
Cross Validation Method	Split
Cross Validation Fold	117
Effect Hierarchy Enforced	None

Number of Observations Read	117
Number of Observations Used	117

Forward Selection Summary				
Step	Effect Entered	Number Effects In	SBC	CV PRESS
0	Intercept	1	10.3714	124.8713
1	x1	2	-6.8675	105.7159
2	x3	3	-24.4842	88.9419
3	x2	4	-26.1273*	85.3803
4	x6	5	-23.0280	85.1732*
* Optimal Value of Criterion				

Selection stopped at a local minimum of the cross validation PRESS.

Stop Details			
Candidate For	Effect	Candidate CV PRESS	Compare CV PRESS
Entry	x5	86.3667 >	85.1732

FIGURE 2.3

SAS codes and LOOCV output for market demand predicted by sale price, competing price, and advertising input

2.4.3   K-fold cross-validation and SAS examples

Although the method LOOCV has many properties in the data learning process, one of the drawbacks of the method is that it is computationally expensive, because the method essentially runs $n - 1$ more time to generate the CVA(n). Another issue with LOOCV is that the data for the trained models are highly correlated since only one observation is removed each time. An extension to LOOCV is the K-fold cross-validation.

Definition 2.12 *K-fold cross-validation: Different from LOOCV, the k-fold cross-validation approach contains the following steps:*

1. *Randomly dividing the original data set into k groups, or k folds of approximately equal size.*

2. *After dividing the data, one of the folds is treated as the validation set and the remaining k-1 folds are treated as the training data to train the model.*

The trained model is evaluated with the held-out one-fold for the MSPE as the model accuracy measurement.

3. *The procedure is repeated k times for each fold serving as the validation set.*

4. *The average of all the k $MSPE_i$ associated with the k-folds, $i = 1, ..., k$ is the overall model accuracy measurement.*

$$CVA(k) = \frac{1}{k} \sum_{i=1}^{k} MSPE_i$$

```
data Fresh;
set WORK.'Fresh_multiple regression data'n;
x1=log('sale price'n);
x2=log('competitor price'n);
x3=log('advertising cost'n);
x4='sale price'n;
x5='competitor price'n;
x6='advertising cost'n;
run;
proc glmselect;
 model 'market demand'n= x1-
x6/selection=forward(stop=CV) details=steps
cvMethod=split(5);
 run;
```

Data Set	WORK.FRESH
Dependent Variable	market demand
Selection Method	Forward
Select Criterion	SBC
Stop Criterion	Cross Validation
Cross Validation Method	Split
Cross Validation Fold	5
Effect Hierarchy Enforced	None

Number of Observations Read	117
Number of Observations Used	117

Forward Selection Summary				
Step	Effect Entered	Number Effects In	SBC	CV PRESS
0	Intercept	1	10.3714	123.7542
1	x1	2	-6.8675	105.9479
2	x3	3	-24.4842	90.0336
3	x2	4	-26.1273*	85.3896
4	x6	5	-23.0280	84.3513*
* Optimal Value of Criterion				

Selection stopped at a local minimum of the cross validation PRESS.

Stop Details			
Candidate For	Effect	Candidate CV PRESS	Compare CV PRESS
Entry	x5	84.7877 >	84.3513

FIGURE 2.4
SAS codes and output for 5-fold CV on market demand data

In what follows, we shall use an example to show the method of cross-validation with SAS in model selection.

Example 2.8 *To seek the relationship between market demand and sale price, advertising cost, and the competitor's price using a random sample of the* *previous months' records as the input data, one of the difficulties is to select the best models among all the possible variables: log scale of the sale price, log scale of the competitor's price, log scale of the advising cost, as well as the original three variables recorded in the dataset. We use the LOOCV and 5-fold cross-validation to select variables for the regression model.*

Examining the SAS outputs in Figure 2.3 and Figure 2.4, we can see that the cross-validation accuracies are consistent, as well as the selected model of the data.

As shown in Figure 2.3, after adding the intercept, the CV PRESS decrease the largest amount (105.72) by adding x_1, log scale of the sale price. The second most important variable in terms of decreasing CV PRESS (cross validation Prediction Residual Error Sum of Squares) is x_3, the log scale of the advertising cost, by the amount of 88.94, which is followed by 85.38 by adding the log scale of the competitor's price into the model. The last variable added to the regression model is x_6, the advertising cost with the CV PRESS value at 85.17. The selection process stops at step 6, where adding the competitor's price into the model results in the increase of CV prediction residual errors sum of squares of the model from 85.17 to 86.37.

$$Demand = \alpha + \beta_1 x_1 + \beta_2 x_3 + \beta_3 x_2 + \beta_4 x6 + \epsilon.$$

Similar conclusions occur when we use the 5-fold cross-validation in Figure 2.4. This example also shows that the log scale transformation of the variable may fit the data better by using cross-validation techniques. We will address the issue of linear regression versus non-linear regression in Chapter 5 and Chapter 6.

2.5 Bootstrapping

Bootstrapping is one of the efficient methods in intrinsic data manipulation. It essentially recovers the population features by repeatedly sampling the original data. Based on the model assumptions in data analytics, there are two types of bootstrapping methods that we shall discuss as a fundamental topic in this section: the nonparametric method and parametric method. The nonparametric method treats the original sample as the population and re-samples the original sample to gain intrinsic data information such as the variance or probability percentiles of the underlying population. The parametric bootstrapping starts with an assumption for the model behind the original sample, uses the original sample to estimate the model parameters, and repeatedly samples the population with estimated parameters for predictions.

2.5.1 Non-parametric bootstrapping

Consider a set of data $E = \{x_1, ..., x_n\}$. When we randomly select n observations (drawing one each time from E with replacement), there are n^n possible different selections. For simplicity, assume that the statistic of interest is the sample mean

$$\bar{X} = \frac{1}{n}\sum_{i=1}^{n} X_i,$$

and the variance of the sample mean

$$Var(\bar{X}) = \sigma^2/n,$$

where the estimate of the population variance σ^2 is the sample variance

$$S^2 = \frac{1}{n-1}\sum_{i=1}^{n}(X_i - \bar{X})^2.$$

Denote $\bar{X}_j^*$ the mean of the jth bootstrapping sample, we have the following results.

Theorem 2.5 *When exhausting all possible n^n re-samples, the mean of the bootstrapping sample mean equals to the mean of the original sample. The sample variance of the bootstrapping sample mean equals to the sample variance of the original sample multiplying by a constant,*

$$c = \frac{n^n(n-1)}{(n^n-1)n^2},$$

and

$$\frac{1}{n^n}\sum_{j=1}^{n^n} \bar{X}_j^* = \bar{X} \tag{2.15}$$

$$\frac{1}{n^n-1}\sum_{j=1}^{n^n}(\bar{X}_j^* - \bar{X})^2 = cS^2. \tag{2.16}$$

Proof: Notice that when we take average over all possible outcomes of the n^n selections with replacement, each element in the original sample E is equally likely to be selected, hence, by regrouping the total entries in the double summation over i and j, so that each summation of n items equals to $\sum_{i=1}^{n} x_i$, which produces n^n terms of $\bar{X}$ in the summation, we have

$$\frac{1}{n^n}\sum_{j=1}^{n^n}\bar{X}_j^* = \frac{1}{n^n}\sum_{j=1}^{n^n}\frac{1}{n}\sum_{i=1}^{n}X_{ij}^*$$

$$= \frac{1}{n^n}\frac{1}{n}\sum_{j=1}^{n^n}\sum_{i=1}^{n}X_{ij}^*$$

$$= \frac{1}{n^n}n^n\bar{X}$$

$$= \bar{X}.$$

This completes the proof of equation (2.15). As for (2.16), with similar rationale in regrouping the re-sampling observations into the n^n items of the original sample, the re-sampling data variance reads,

$$\frac{1}{n^n-1}\sum_{j=1}^{n^n}(\bar{X}_j^* - \bar{X})^2$$

$$= \frac{1}{n^n-1}\sum_{j=1}^{n^n}(\frac{1}{n^2}[\sum_{i=1}^{n}(x_{ij}^* - \bar{X})]^2)$$

$$= \frac{1}{n^n-1}\sum_{j=1}^{n^n}(\frac{1}{n^2}[\sum_{i=1}^{n}(x_{ij}^* - \bar{X})^2 + \sum_{i\neq k}(x_{ij}^* - \bar{X})(x_{kj}^* - \bar{X})])$$

$$= \frac{1}{n^n-1}\sum_{j=1}^{n^n}(\frac{1}{n^2}[\sum_{i=1}^{n}(x_{ij}^* - \bar{X})^2])$$

$$= \frac{1}{n^n-1}\frac{1}{n^2}\sum_{j=1}^{n^n}\sum_{i=1}^{n}(x_{ij}^* - \bar{X})^2$$

$$= \frac{1}{n^n-1}\frac{1}{n^2}(n-1)n^n S^2$$

This completes the proof of (2.16).

Theorem 2.5 establishes the connection between the mean and sample variance of the bootstrapping sample means under the setting of equally likely selection. The following theorem sets connection between the average of n^n bootstrapping samples and the sample mean of the original sample.

Theorem 2.6 *Let S_j^{2*} be the sample variance of the bootstrapping sample $\{x_{1j}^*, ..., x_{nj}^*\}$. We have*

$$\frac{1}{n^n}\sum_{j=1}^{n^n}S_j^{2*} = \frac{n-1}{n}S^2. \tag{2.17}$$

2	3	7	4	2.645751311	7	Original data

Re-sampling information

possible re-sampling outcomes			mean	std	variance
2	2	2	2	0	0
2	2	3	2.333333	0.577350269	0.333333
2	2	7	3.666667	2.886751346	8.333333
2	3	2	2.333333	0.577350269	0.333333
2	3	3	2.666667	0.577350269	0.333333
2	3	7	4	2.645751311	7
2	7	2	3.666667	2.886751346	8.333333
2	7	3	4	2.645751311	7
2	7	7	5.333333	2.886751346	8.333333
3	2	2	2.333333	0.577350269	0.333333
3	2	3	2.666667	0.577350269	0.333333
3	2	7	4	2.645751311	7
3	3	2	2.666667	0.577350269	0.333333
3	3	3	3	0	0
3	3	7	4.333333	2.309401077	5.333333
3	7	2	4	2.645751311	7
3	7	3	4.333333	2.309401077	5.333333
3	7	7	5.666667	2.309401077	5.333333
7	2	2	3.666667	2.886751346	8.333333
7	2	3	4	2.645751311	7
7	2	7	5.333333	2.886751346	8.333333
7	3	2	4	2.645751311	7
7	3	3	4.333333	2.309401077	5.333333
7	3	7	5.666667	2.309401077	5.333333
7	7	2	5.333333	2.886751346	8.333333
7	7	3	5.666667	2.309401077	5.333333
7	7	7	7	0	0

4   grant mean

1.615385   grant variance

4.666667   mean variance

7   mean-variance*n/(n-1)

7   grant-vatriance*[(n^n)-1]*n^2/[(n^n)*(n-1)]

FIGURE 2.5

Small sample nonparametric bootstrap

Proof. The left-hand side of (2.17) reads

$$\frac{1}{n^n} \sum_{j=1}^{n^n} S_j^{2*}$$

$$= \frac{1}{n^n} \sum_{j=1}^{n^n} \frac{1}{n-1} \sum_{i=1}^{n} (x_{ij}^* - \bar{X}_j^*)^2$$

$$= \frac{1}{n^n} \sum_{j=1}^{n^n} \frac{1}{n-1} \sum_{i=1}^{n} (x_{ij}^* - \bar{X} + \bar{X} - \bar{X}_j^*)^2.$$

Decomposing the terms above reads,

$$\frac{1}{n^n}\sum_{j=1}^{n^n} S_j^{2*}$$

$$= \frac{1}{n^n}\sum_{j=1}^{n^n}\frac{1}{n-1}[\sum_{i=1}^{n}(x_{ij}^* - \bar{X})^2 - 2\sum_{i=1}^{n}(x_{ij}^* - \bar{X})(\bar{X} - \bar{X}_j^*) + \sum_{i=1}^{n}(\bar{X} - \bar{X}_j^*)^2$$

$$= \frac{1}{n^n}\sum_{j=1}^{n^n}\frac{1}{n-1}\sum_{i=1}^{n}(x_{ij}^* - \bar{X})^2 - \frac{1}{n^n}\sum_{j=1}^{n^n}\frac{n}{n-1}(\bar{X} - \bar{X}_j^*)^2$$

$$= S^2 - \frac{1}{n^n}\frac{n}{n-1}\frac{n^n(n-1)}{n^2}S^2$$

$$= \frac{n-1}{n}S^2,$$

by (2.16). This completes the proof of Theorem 2.6.

The following numerical example illustrates the above two theorems on nonparametric bootstrapping methods when $n = 3$.

Example 2.9 *Consider the original data set $E = \{2, 3, 7\}$, as show in Figure 2.5, the original sample has sample mean 4 and sample variance 7. We have $3^3 = 27$ different sample values in the re-sampling with replacement method. Taking average of all the 27 re-sampling means, we get the exactly same value as the sample mean of the original sample.*

As shown in Figure 2.5, when we take the sample variance of the bootstrapping sample means, the grant variance is 1.615385, which reaches the original sample variance 7 after multiplying by

$$\frac{(n^n - 1)n^2}{n^n(n-1)} = \frac{26*9}{27*2}.$$

As for theorem 2.6, the mean (average) of the bootstrapping samples is 4.666667, which comes back to the original sample variance 7 after multiplying by

$$\frac{n}{n-1} = \frac{3}{2}.$$

2.5.2 Parametric bootstrapping

When there is an assumption on the underlying model for the original sample, the additional model condition should be taken into consideration in the generation of the re-sampling process. In this case, the model parameters will be estimated with the original sample. Once the parameters are estimated, the underlying model for the original sample can be used to add information toward the re-sampling data.

We will use an example to demonstrate the difference between parametric bootstrapping and non-parameter bootstrapping.

```
> set.seed(10)
> x<-rnorm(20, 2, 0.8)
> x
 [1] 2.0149969 1.8525980 0.9029356 1.5206658 2.2356361 2.3118354
1.0335391 1.7090592 0.6986619 1.7948173 2.8814236
[12] 2.6046252 1.8094132 2.7899558 2.5931121 2.0714778 1.2360449
1.8438797 2.7404170 2.3863828
>
> x.mean <- mean(x)
> x.std <-sd(x)
>
> x.mean
[1] 1.951574
> x.std
[1] 0.6399275


Nonparametric bootstrapping

> vec<-rep(0, 10000)
> for (i in (1:10000)){
+    y<-sample(x, 20, replace=TRUE)
+    vec[i]<-mean(y)
+ }
> mean(vec)
[1] 1.951677
> sd(vec)
[1] 0.1416576
> sqrt(20)*sd(vec)
[1] 0.6335121


Parametric bootstrapping with normal model assumption

> for (i in (1:10000)){
+      y<-rnorm(20, x.mean,x.std)
+      vec[i]<-mean(y)
+    }
> mean(vec)
[1] 1.949943

> sd(vec)
[1] 0.1415905

> sqrt(20)*sd(vec)
[1] 0.633212
```

FIGURE 2.6
Comparing nonparametric and parametric bootstrapping

Example 2.10 *Assume that the original sample contains 20 observations generated from a normal model with mean=2 and standard deviation=0.8. With the original data, we can run non-parametric bootstrapping to estimate the unknown mean and the unknown standard deviation. As shown in Figure 2.6, the bootstrapping sample mean is 1.9517 and the standard deviation as 0.6335. The non-parameter bootstrapping results are very close to the sample mean and sample standard deviation of the original one, 1.952 for the sample mean and 0.6399 for the sample standard deviation.*

When the parametric bootstrapping is used, the re-sampling data are now generated from a normal model with mean 1.952 and standard deviation 0.6399. The re-sampling data now are not drawn from the original sample. Instead, they are drawn from a normal model with mean 1.952 and standard deviation 0.6399. As shown in Figure 2.6, the mean value of the re-sampling data is not 1.9499 with standard deviation 0.6332. These results are very close to the outcomes using the non-parametric method.

Certainly, in Example 2.9 and Example 2.10, the estimated value is the unknown mean and the corresponding variance is the sample standard deviation. Obviously, the bootstrapping method does not add much information on the predicted outcome from the original sample. In the following example, we explore a situation where the original data sample does not provide an estimation of the sample standard deviation of the parameter estimator. In this case, bootstrapping becomes an effective way to find the standard deviation, hence the confidence interval.

Example 2.11 *Blood pressure instability is a critical issue in the analysis of treatment regime for hypertension patients. As an example, consider the treatment regime with three medications Losartan, Valsartan, and Bisoprolol, attributing to the drug efficacy. The measurement for blood pressure fluctuation becomes*

$$Z = \alpha L + \alpha V + (1 - 2\alpha)B + c, \tag{2.18}$$

where c is the constant of the patient's baseline blood pressure. Since both Losartan and Valsartan target the Angiotensin II receptor blockers while Bisoprolol targets Beta-blockers. The question is to find the proper weight of α, the right proportion for the Angiotensin II receptor blockers, in the treatment regime so that the blood pressure variation reaches the minimum possible level. Here, the effects of Losartan and Valsartan are correlated since they target the same receptor blockers, but the impact of Bisoprolol is not correlated to Losartan or Valsartan because it targets different receptor blockers.

Solution: According to (2.18), the blood pressure variability reads

$$
\begin{aligned}
Var(Z) &= Var(\alpha L + \alpha V + (1 - 2\alpha)B + c \\
&= \alpha^2 Var(L) + \alpha^2 Var(V) + (1 - 2\alpha)^2 Var(B) + 2\alpha^2 Cov(L, V),
\end{aligned}
$$

since effects of Losartan and Valsartan are correlated, but the impact of Bisoprolol is not correlated to Losartan or Valsartan, we have

$$Cov(L, V) \neq 0 \quad Cov(L, B) = Cov(V, B) = 0.$$

Thus, the blood pressure fluctuation is a function of the medication proportion α in the construction of treatment regime when the drug effect variations $V(L)$, $V(S)$, $V(B)$, and correlations $Cov(L, S)$, are pre-determined in the stage of drug development (prior to the study of treatment regime). This leads to the conclusion that, in this study of treatment regime, α is the only factor attributing to the concern of blood pressure instability.

$$f(\alpha) = \alpha^2 Var(L) + \alpha^2 Var(V) + (1 - 2\alpha)^2 Var(B) + 2\alpha^2 Cov(L, V). \tag{2.19}$$

Taking derivative of $f(\alpha)$ in (2.19) with respect to α gets, after some algebra calculation,

$$\hat{\alpha} = \frac{2Var(B)}{Var(L) + Var(S) + 4Var(B) + 2Cov(L,S)}. \tag{2.20}$$

With a set of samples on the fluctuation of blood pressure records from clinical trials, we can use plug in moment estimation method to estimate the treatment regime proportion $\hat{\alpha}$, however, the estimation of the standard deviation of $\hat{\alpha}$ is not available. Under this scenario, the bootstrapping method provides a convenient and efficient alternative.

```
> library(boot)
> bp <- read.table("D:/chapter2/bp.txt", header=TRUE)
> head(bp)

    Losartan  Valsartan  Bisoprolol
1 -1.5453679 -1.1720908 -0.37967880
2  2.0408970  1.4170437  2.27528290
3 -0.1547983 -0.1231551  2.01227522
4  0.8056872  0.4575227 -0.03546461
5 -1.0266717 -0.2783768 -0.86678710
6  2.1577671  1.1213169  0.68332724

> alpha.fn <-function(data, index){
+    X1 <-data$Losartan[index]
+    X2 <- data$Valsartan[index]
+    X3 <- data$Bisoprolol[index]
+    (2*var(X3))/(var(X1)+var(X2)+4*var(X3)+2*cov(X1,X2))
+ }

> (alp <- alpha.fn(bp, 1:200))
[1] 0.2970361
> boot1<-boot(bp, alpha.fn, R=1000)
> alpha.CI.upper <- alp + 1.96*apply(boot1$t,2,sd)[1]
> alpha.CI.lower <- alp - 1.96*apply(boot1$t,2,sd)[1]
> alpha.CI.lower
[1] 0.263982
> alpha.CI.upper
[1] 0.3300902
```

FIGURE 2.7
Optimal treatment regime for blood pressure instability

As depicted in Figure 2.7, with the original sample of blood pressure fluctuation of 200 patients in the clinical trial, some patients responded positively to medications on Beta-blockers, some on Angiotensin II receptor blockers, and some on both of them. With the original sample, plugging-in the moment estimators of the variations and covariation in (2.20) gets

$$\hat{\alpha} = 0.297,$$

which means that if 29.7% of the hypertension medication on Angiotensin II receptor blockers and 70.3% on Beta-blockers, the drug treatment regime reaches its optimal level in keeping the blood pressure stable for the patients.

However, the data is unable to estimate the standard deviation of $\hat{\alpha}$ for the 95% confidence range on the optimal treatment regime. Figure 2.7 shows that with the bootstrapping method and 1000 re-sampling data, we have the bootstrap mean,

$$\bar{\alpha} = \frac{1}{1000} \sum_{r=1}^{1000} \hat{\alpha}_r = 0.2970361.$$

And the bootstrap standard deviation,

$$\sqrt{\frac{1}{1000-1} \sum_{r=1}^{1000} (\hat{\alpha}_r - \bar{\alpha})^2} = 0.01686434,$$

which leads to a 95% confidence level $(0.264, 0.33)$ for the unknown optimal proportion α. This means that the optimal treatment regime locates in the range from 26.4% to 33% for the best blood pressure stability of hypertension patients.

Summary

This chapter focuses on basic concepts and methods that facilitate follow-up discussions on statistical prediction and machine learning. It starts with a discussion on different types of data, which is the first step in the learning process. Different types of data require different methods of measure for homogeneity and learning procedures. This is often overlooked in machine learning in practice. If the method is not right, the trained model could be fatally misleading, even if it reaches a small testing error in *one* testing dataset. We used case-control data, cohort data, and cross-sectional data in this chapter to elucidate discernible methods and outcomes corresponding to the data.

The second key component that we concentrate in this chapter is decision tree, a concept that we will frequently use and intertwine with other topics in the rest of the book. We introduce the mathematical definition and practical interpretation of a decision tree, which changes the conventional inference approach on the culture camp of model-based data science. More insightful issues in this regard will be delineated in Chapter 9.

Similar to the fundamental concepts on the probability of type-I error and the probability of type-II error in hypothesis testing, another frequently used terminology in the data-oriented camp is the concepts of sensitivity and specificity, with the plot of the two measurements by the ROC curve. We enhance the definition with introductory examples in this chapter and will explore further on this topic regarding the trade-off between sensitivity-specificity in Chapter 3.

Cross-validation and bootstrapping are two data-based computer-intensive methods in the data-oriented culture camp. We introduce them in this chapter with a theoretical discussion on LOOCV for linear models and an illustrating example on k-fold cross-validation in SAS. It should be mentioned that these two methods are frequently intertwined with other data science methods in the rest of the book, such as linear prediction (Chapter 4), non-linear prediction (Chapter 5), support vector machine (Chapter 8), and range regression (Chapter 9). More intrinsic discussion on bootstrapping methods can be found in Efron (1979) [47], Efron and Tibshirani (1993) [48], and Davision and Hinkley (2006) [41], among others.

3

Sensitivity and Specificity Trade-off

For problems involving classification or disease prediction, we are often tasked with making a decision for a binary response. For example, diagnosing healthy or diseased subjects, detecting male or female persons in facial recognition, etc. Making decisions for binary outcomes always involves two errors: the error of incorrectly diagnosing a disease, $P(claiming\ disease\ |healthy)$, and the error of missing a disease, $P(claiming\ health\ |case)$.

If *we assert every subject as healthy*, we will never mistakenly diagnose a patient,

$$P(claiming\ disease\ |healthy) = 0,$$

but we will surely commit an error missing cases in the population,

$$P(claiming\ health\ |case) = 1.$$

On the other hand, if we assert everyone as having the disease (do not claim any body as being healthy),

$$P(claiming\ health\ |case) = 0,$$

consequently, we will completely misdiagnose healthy subjects in the study,

$$P(claiming\ disease\ |healthy) = 1.$$

This behavior is known as the trade-off between sensitivity and specificity. In hypothesis testing, this statistical concept is closely related to the trade-off between the probability of making type-I and the probability of making type-II error. In the following section, we will discuss the trade-off in detail and address new methods resolving the dilemma regarding the sensitivity-specificity trade-off.

3.1   Dilemma on false positive and false negative errors

As discussed in Chapter 2, when we predict a binary outcome (such as diseased vs healthy) with diagnostic measure D, assuming that a low value of D is associated with the disease under investigation, the concepts of sensitivity and specificity are defined as

$$Sensitivity = P(D < c|case) \quad Specificity = P(D \geq c|control), \qquad (3.1)$$

where c is the diagnostic criterion. The diagnostic measure D may be derived from a logistic regression model, or from clinical results. For example, when we use complete blood count (CBC) to diagnose leukemia, CBC measures the number of red blood cells, white blood cells, and platelets in a patient's blood. It also provides information on the amount of hemoglobin (oxygen carriers) and hematocrit (proportion of red blood cells in the blood). Although a low value of CBC is linked to leukemia, CBC changes alongside demographic features such as age, gender, race, BMI, and comorbidity. CBC can even change from time to time (longitudinal effect) within the same individual. Thus, it is unrealistic to pinpoint a cut-off value of the diagnostic threshold c in diagnosing leukemia. For a specific population, the threshold c needs to be calculated from the training data in disease prediction.

On one hand, if c is set too high, more patients will have their CBC measure D value satisfying the condition $D < c$. This may result in more patients being incorrectly classified into the diseased group. Especially, misdiagnosing healthy subjects as having leukemia, incurs the false positive error. On the other hand, if the threshold c is set too low, less people will have their CBC test result satisfying the condition $D < c$, and more people will have their CBC test result satisfying $D \geq c$. This leads to more people being incorrectly classified as being healthy, causing false negative error.

Theoretically, one may always set the criterion c to be higher than the highest possible CBC value to completely avoid the false positive error, since such c values result in the claim that everyone is positive. However, this leads to the case where sensitivity=1 and specificity=0 as pointed out in (3.1). Alternatively, when c is set to any value below the lowest possible CBC value, such c value leads to the assertion that everyone is healthy, resulting in sensitivity=0 and specificity=1 by (3.1). In general, for the scenario of predicting leukemia using CBC, within the permissible range of threshold c defined in (3.1), higher value of c increases the sensitivity but decreases the specificity, and lower value of c increases the specificity but decreases the sensitivity. As such, there is a dilemma in the control of sensitivity and specificity in the determination of the threshold c value.

In logistic regression analysis with a binary response variable, the threshold c is, in general, selected in the following way. First, all permissible values of c are used to construct an estimated ROC curve by plotting the following pairs of points

$$(1 - specificity(c), \ sensitivity(c)) \quad c \in A,$$

where A is the set of all permissible c values. The area under the ROC curve is typically used to select the prediction model, and the value of c is adjusted to produce a sensitivity and specificity with closest distance to the optimal point in the ROC plot, where sensitivity=specificity=1. Namely,

TABLE 3.1

Type-I and Type-II errors in hypothesis testing

	Rejecting null hypothesis	Not rejecting null hypothesis
Null true	Type-I error	correct decision
Alternative true	power	Type-II error

$$c = ArgInf_{c \in A}||d(c)||, \tag{3.2}$$

where

$$||d(c)|| = \sqrt{(1 - specificity(c))^2 + (1 - sensitivity)^2}$$
$$= \sqrt{(false\ positive\ error(c))^2 + (false\ negative\ error)^2}. \tag{3.3}$$

It should be noted that (3.3) is only one of the selection approaches for the diagnostic criterion c. It uses the convenient concept of the distance between two points in the xy-plane. This selection is not necessarily the most optimal choice in practice. For example, when controlling for the false negative error is more important, such as misdiagnosing and missing treatments of a life threatening disease for a patient versus the error of asking the patient to take a second confirmatory screening test, the selection standard in (3.3) will be misleading, because it did not take the weights on diagnostic priority into consideration. On the other hand, when controlling false positive error is more critical, such as the error of incorrectly pushing a healthy person into an operation room versus the error of asking the patient to take a preventive medicine, the selection standard in (3.3) is also misleading, because it treats both errors at the same level of importance (equal weights).

Controlling false positive error and false negative error is not a new paradox to data science. In hypothesis testing, we are often confronted with the dilemma of controlling the probability of Type-I error (incorrectly rejecting the null hypothesis) and the probability of Type-II error (incorrectly rejecting the true alternative hypothesis) as shown in Table 3.1.

According to well-documented statistics literature, the solution to the dilemma on the control of the Type-I and Type-II errors is to select one statement as the null hypothesis, control the probability of making the Type-I error in the selection of rejection areas, and find the most powerful test to minimize the chance of making the Type-II error.

In what follows in this chapter, we will reformulate the control of false positive error and the control of false negative error into the framework of hypothesis testing, and consequently state similar results in the determination of the diagnostic threshold by keeping the control on false positive rate, and minimizing the chance of making the false negative rate. The idea is similar to the concept of uniformly most powerful test in hypothesis testing.

3.2 Most sensitive diagnostic variable

We shall introduce a new concept in the evaluation of a diagnostic procedure, the *Uniformly Most Efficient* predictor, which is equivalent to a global optimization solution when the target function aims to maximize the sensitivity of the diagnostic measure given a pre-specified level of specificity. Notice that, as defined in (3.1), both sensitivity and specificity depend on the diagnostic measure D and diagnostic threshold c.

Definition 3.1 Uniformly Most Efficient Lower Variable: *Assume that lower values of the diagnostic measure are associated with the disease. Let K be a set of diagnostic measures that satisfy*

$$P(D \geq c | healthy) = 1 - \alpha,$$

for a pre-specified level of specificity $1 - \alpha$. The most efficient lower variable refers to the diagnostic measure D^* that satisfies the following conditions.

I) $D^* \in K$, which means $P(D^* \geq c^* | healthy) = 1 - \alpha$.

II) $P(D^* < c^* | case) \geq P(D < c | case)$ for any $D \in K$, which means that D^* is the one with the highest sensitivity among all the diagnostic measures that have specificity $1 - \alpha$.

When high values of the diagnostic measure D are associated with the disease, such as escalated systolic blood pressure or escalated cholesterol level for heart attacks, Definition 3.1 is equivalent to the following.

Definition 3.2 Uniformly Most Efficient Upper Variable: *Assume that high values of the diagnostic measure are associated with the disease. Let K be a set of diagnostic measures that satisfy*

$$P(D < c | healthy) = 1 - \alpha,$$

for a pre-specified level of specificity $1 - \alpha$. The most efficient upper variable refers to the diagnostic measure D^* that satisfies the following conditions.

1 $D^* \in K$, which means $P(D^* < c^* | healthy) = 1 - \alpha$.

2 $P(D^* \geq c^* | case) \geq P(D \geq c | case)$ for any $D \in K$, which means that D^* is the one with the highest sensitivity among all the diagnostic measures that have specificity $1 - \alpha$.

Definition 3.1 and Definition 3.2 are essentially the same except the direction of the diagnostic measure toward the disease. Notice that with the two evaluation criteria (false positive and false negative errors, or type-I and type-II errors), the traditional approach is to fix one evaluation criterion (significance level) and optimize the second one (power). While for the definition of uniformly most efficient measures, we essentially fix the probability of the false negative error (specificity), and find the largest possible sensitivity. In what follows, we shall use a simulation example to obtain a better understanding of definition 3.1 and definition 3.2.

Example 3.1 *Low-density lipoprotein cholesterol (LDL-C) level is a risk factor for coronary heart disease. Assume that the LDL-C level of healthy population follows a normal model with mean 130 mg/dL and a standard deviation 3 mg/dL. Also assume that LDL-C levels of patients with coronary heart disease follow a normal model with mean μ more than 130 mg/dL and the same variation (standard deviation). If we have the blood test results of 20 subjects at the similar LDL-C levels, we want to identify the diagnostic measure for coronary heart disease patients so that the specificity is kept at 95% level.*

Solution Set the level $\alpha = 0.05$ so that the specificity is at 0.95 level. We have

$$Specificity = P(\{\mathbf{X} : \frac{\bar{X} - 2}{3/\sqrt{20}} < z_\alpha | healthy\} = 0.95,$$

where $\mathbf{X}$ is the random sample, $\bar{X}$ is the sample mean, and z_α is defined as $P(Z > z_\alpha) = \alpha$, in which Z follows the standard normal distribution.

Thus, for the diagnostic predictor $D = \bar{X}$, the threshold for 0.95 specificity reads,

$$c = 130 + 1.645 * 3/\sqrt{20}.$$

Under this setting, the sensitivity becomes

$$\begin{aligned} Sensitivity(\mu) &= P(\frac{\bar{X} - 130}{3/\sqrt{20}} > z_\alpha | \mu) \\ &= P(\frac{\bar{X} - \mu}{3/\sqrt{20}} > z_\alpha - \frac{\mu - 130}{3/\sqrt{20}}) \\ &= P(Z > z_\alpha - \frac{\mu - 130}{3/\sqrt{20}}). \end{aligned}$$

Notice that in this example, the sensitivity is a monotonic increasing function with mean LDL-C level μ.

We may run the process 10,000 times to examine the sensitivity as a function of the LDL-C level. As shown in Figure 3.1, when the patient LDL-cholesterol level increases, the sensitivity associated with the diagnostic measure D increases. The corresponding R-code is also included in Figure 3.2.

Example 3.1 shows the approach to find the diagnostic measure so that the specificity can satisfy a given level. However, it does not prove whether the diagnostic measure is a uniformly most efficient predictor. The following optimal criterion ensures that the diagnostic measure D in Example 3.1 is

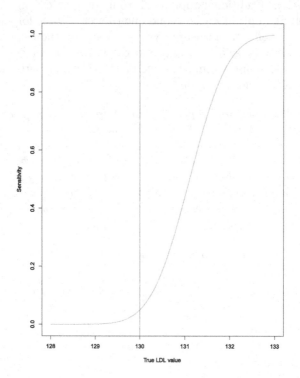

FIGURE 3.1
Sensitivity in patients with LDL-C level more than 130 mg/dL

a most efficient predictor. In general, when the underlying models can be plausibly assumed, we use the likelihood ratio criterion to seek the uniformly most efficient diagnostic predictor.

Theorem 3.1 *Assume that the underlying model (pdf or pmf) of a set of data is $f_1(x)$ for case, and $f_0(x)$ for control. The most efficient diagnostic measure is*

$$D^* = \frac{f_1(x)}{f_0(x)}$$

with

$$P(D^* < c^* | healthy) = 1 - \alpha$$

when small diagnostic measure D is associated with the disease, and

$$P(D^* > c^* | healthy) = 1 - \alpha$$

when large diagnostic measure D is associated with the disease.

```
sigma <-3

n <-20

mu0 <-130

mu.o<-seq(128, 133, 0.03)

POWER <-matrix(0, length(mu.o),2)

result <- 0
for (i in 1:length(mu.o)){
  mu<-mu.o[i]
    result <- 1-pnorm(qnorm(0.95)-(mu-130)/(3/sqrt(20)))
   POWER[i,] <-c(mu, result)
}
png(file="~/desktop/saving_plot2.png",
  width=500, height=400)

plot(POWER, xlim=c(128, 133), xlab="True LDL value",
ylab="Sensitivity", type="l", col="green", lty=1, lwd=2)
abline(v=130, col="red")
```

FIGURE 3.2
Code for sensitivity in patients with LDL-C level more than 130 mg/dL

Note that the above theorem provides an approach to find the most efficient diagnostic predictor when the likelihood function of the disease and the healthy population can be plausibly assumed.

Proof *Assume that large values of diagnostic measure D are associated with the disease. For any diagnostic measure D with*

$$P(D > c|healthy) = 1 - \alpha,$$

denote

$$A = \{D > c\} \quad A^* = \{D^* > c^*\},$$

we have

$$(I_A - I_{A^*})[D^*(X) - c^*] \leq 0,$$

which is equivalent to

$$(I_A - I_{A^*})[f_1(X) - c^* f_0(X)] \leq 0,$$

hence

$$\int (I_A - I_{A^*})[f_1(X) - c^* f_0(X)]dX \leq 0$$

where $f_i(X) = \prod_{j=1}^{n} f(x_j|\theta_i)$ *for* $i = 0, 1$. *Now*

$$\int (I_A - I_{A^*})f_1(X)dX \leq c^* \int (I_A - I_{A^*})f_0(X)dX.$$

The right-hand side is non-positive since

$$P(D > c|healthy) = specificity = 1 - \alpha,$$

and

$$P(D^* > c^*|healthy) = specificity = 1 - \alpha.$$

We have

$$\int (I_A - I_{A^*})f_1(X)dX \leq 0.$$

This implies that

$$\int I_A f_1(X)dX \leq \int I_{A^*} f_1(X)dX.$$

$$P(A|case) \leq P(A^*|case)$$

This proves that D^* is the most efficient diagnostic measure and c^* is the most efficient diagnostic threshold.

   When small values of diagnostic measure D are associated with the disease, similar argument completes the proof of the theorem.

   In the following examples, we present two scenarios (one for continuous model and one for discrete model) to illustrate the application of Theorem 3.1 in the construction of the most efficient diagnostic predictor.

Example 3.2 *Assume that the LDL-cholesterol levels follow a normal model with $X_i \sim N(130, 2)$ for healthy subjects and $X_i \sim N(150, 2)$ for patients with coronary heart diseases. If the specificity is set to 0.95, we want to find the most efficient diagnostic predictor.*

Solution: In this case, the two possible models are

$$f(X|healthy) = (\frac{1}{\sqrt{2\pi}\sigma})^n \exp\{-\frac{1}{2\sigma^2} \sum_{i=1}^{n}(x_i - 130)^2\}$$

and

$$f(X|case) = (\frac{1}{\sqrt{2\pi}\sigma})^n \exp\{-\frac{1}{2\sigma^2} \sum_{i=1}^{n}(x_i - 150)^2\}.$$

By Theorem 3.1, the most efficient diagnostic measure reads,

$$D(X) = \frac{f(X|case)}{f(X|healthy)}.$$

Since the escalated LDL-Cholesterol level is associated with coronary heart diseases, the most efficient diagnostic predictor is

$$\{D(X) > c\} = \{X : \sum_{i=1}^{n}(x_i - 5)^2 < \sum_{i=1}^{n}(x_i - 1)^2 + k_\alpha^*\},$$

for a constant k_α^* that depends on the specificity level $1 - \alpha$, and

$$\{D(X) > c\} = \{X : \overline{X} > k_\alpha^{**}\}.$$

Now, if all the patients are healthy, the sample mean statistic of their LDL-Cholesterol levels follows $N(130, \frac{2}{\sqrt{n}})$. After standardizing the sample statistic gets

$$\{D(X) > c\} = \{X : \frac{\overline{X} - 130}{\frac{2}{\sqrt{n}}} > k_{0.05}^{***}\}.$$

Notice that the evaluation criterion requires that the specificity is 0.95,

$$P(D(X) < c|healthy) = 0.95,$$

thus, we have $k_{0.05}^{***} = 1.645$. The most efficient diagnostic predictor, for any sample size n, reads

$$\{D^* > c^*\} = \{X : \frac{\overline{X} - 130}{\frac{2}{\sqrt{n}}} > 1.645\}.$$

In the next example, we shall discuss an example of finding the most efficient diagnostic predictor in a discrete model for the diagnosis of Type-I diabetes.

Example 3.3 *Let X be a random variable associated with diabetes symptoms, including numbness, weight loss, swollen gums, slow healing, increased appetite, blurred vision, energy loss, and frequent thirst. Assume that from historical data of patient records, the chance of each symptom for healthy patients (blood glucose level ≤ 100) versus diabetes patients (blood glucose level > 100) in a local hospital are given in the following table.*

x	numb	weight loss	swollen gums	slow heal	increase appetite	blur vision	energy loss	often thirst
$f(x|H)$	.02	.02	.02	.01	.05	.01	0.41	0.46
$f(x|D)$	.22	.02	.12	.04	.18	.02	.06	0.34

Based on the symptoms of a newly admitted patient, the hospital is interested in diagnosing whether the patient has diabetes with a requirement that the specificity should be at least 95%.

Solution: Notice that in this case, the ratio of chances for each symptom $D(X) = \frac{f(x|case)}{f(x|healthy)}$ takes the following values

x	numb	weight loss	swollen gums	slow heal	increase appetite	blur vision	energy loss	often thirst
$f(x|H)$	0.02	0.02	0.02	0.01	0.05	0.01	0.41	0.46
$f(x|D)$	0.22	0.02	0.12	0.04	0.18	0.02	0.06	0.34
λ	11	1	6	4	3.6	2	0.146	0.739

In this case, a higher ratio of chance indicates that the individual is more likely to have the disease. Based on Theorem 3.1 and according to the ranking of the diagnostic measure for each symptom, we arrange the symptoms by the likelihood of diabetes verse diabetes. To satisfy the evaluation criterion of controlling the rate of misdiagnosis at 5% level, namely

$$Specificity = P(D < c|healthy) = 0.95,$$

symptoms with likelihood ratios on the top 5% misdiagnostic rate, are

$$\{D^* > c^*\} = \{numbness, swollen\ gums, slow\ healing\}.$$

The chance of correctly diagnosing a diabetes patient, which is the sensitivity of this diagnostic predictor reads

$$Sensitivity = P(D^* > c^*|case) = 0.22 + 0.12 + 0.04 = 38\%.$$

Certainly, when the blood glucose level test is available, the laboratory test result is more accurate in detecting diabetes as a follow-up diagnosis. However, as illustrated by Theorem 3.1, this example shows that the selection of diagnostic predictors is possible without the use of the continuous likelihood function.

It should be noted that Example 3.3 and Example 1.5 are very similar in a way where the control of the type-I error in hypothesis testing plays the same role as the control of the false negative error in the sensitivity-specificity analysis.

In the above two examples, predictions are made on an unknown parameter for assumed models. The intuition behind it relies on the likelihood ratio that optimizes the evaluation criterion by maximizing the sensitivity of the diagnostic predictor while controlling the specificity. The rationale of theorem 3.1 is grounded on the intuition that we diagnose the patient as sick when the likelihood of sickness is higher compared with the likelihood of being healthy. This idea consequently leads to the likelihood ratio measurement in the construction of the most efficient diagnostic predictor.

Definition 3.3 Likelihood Ratio measurement: *Assume that patient observations follows the model $l_1(\theta, X)$ for disease population and $l_2(\theta, X)$ for healthy population. The likelihood ratio measurement is defined as*

$$D(X) = \frac{\sup_{\Theta_0} L(\theta|X)}{\sup_{\Theta} L(\theta|X)}$$

where Θ_0 is the set for parameters in the likelihood of the healthy population, Θ_1 the disease population, and $\Theta = \Theta_0 \bigcup \Theta_1$ is the whole parameter space.

Heuristic rationale of the *likelihood ratio measurement* for diagnostic prediction: When the underlying model of the data is assumed to represent the patient features, if the likelihood ratio is small, the largest possible likelihood that the patient in the healthy population is less than the largest possible likelihood that the patient is sick. Thus, we diagnose the patient as having the disease. The key idea in this model-based sensitivity-specificity analysis is the assumption of the underlying model. If the assumption is invalidated, the diagnostic predictor consequently becomes invalid in the search for the most efficient one.

Example 3.2 defines the healthy population's LDL-Cholesterol level equal to 130mg/dL and sick population at 150mg/dL. However, nobody in practice has an LDL-Cholesterol level exactly equal to those two specific numbers, although the definition is mathematically sound. With the assumed underlying model, we shall consider scenarios where the definition of sickness is extended to LDL-Cholesterol being more than 130 mg/dL, and the definition of a healthy population is extended to the corresponding level being less than 130mg/dL. We use the following example to show how to seek the most efficient diagnostic predictor with the likelihood ratio measurement.

Example 3.4 *Let $X_1......X_n$ be a random sample of blood test LDL-cholesterol readings. Assume that the readings follow a population distribution $N(\mu, 1)$ with an unknown common mean LDL-Cholesterol level, μ. We are interested in finding the most efficient diagnostic predictor to diagnose whether the population is sick (μ more than 130) or healthy (μ less than 130) with the specificity at 95% level.*

Solution: Since the specificity is set at 95%, we have

$$\sup_{\mu \leq 130} P(D > c|Healthy) = 0.05.$$

On the other hand, for the sensitivity across all LDL-Cholesterol level more than 130, we need,

$$P(D > c|case) = P(\bar{X} > c^*|case) = P\left(Z > \frac{c^* - \mu}{1/\sqrt{n}}|\mu > 130\right).$$

For any two LDL-Cholesterol levels λ_1 and λ_2, if $\lambda_1 \leq 130$ and $\lambda_2 > 130$, by the derivation in Example 3.2, the most efficient diagnostic predictor is

$$\{D > c\} = \{\bar{X} > \lambda_1 + 1.645 * \frac{1}{\sqrt{n}}\}$$

for any value λ_1 and λ_2. Thus, the difficulty becomes to find the optimal value c in the supremum. For notational convenience, denote μ=LDL-cholesterol level. We need to find,

$$\sup_{\mu \leq 130} P(Z > \frac{c^* - \mu}{1/\sqrt{n}} | healthy).$$

Notice that

$$P(Z > \frac{c^* - \mu}{1/\sqrt{n}} | control) = P(Z > \frac{c^* - 130 + 130 - \mu}{1/\sqrt{n}} |)$$

$$\leq P(Z > \frac{c^* - 130}{1/\sqrt{n}} | \mu \leq 1)$$

$$= P(Z > \frac{c^* - 130}{1/\sqrt{n}}) = 0.05.$$

So, setting $\frac{c^* - 130}{1/\sqrt{n}} = 1.645$ gets $c^* = 130 + 1.645 \frac{1}{\sqrt{n}}$. This leads to the most efficient predictor, D,

$$\{D > c\} = \{\mathbf{x} : \bar{X} > 130 + 1.645\sqrt{n}\}.$$

We have discussed two approaches, the likelihood ratio measurement and Theorem 3.1 for the construction of the uniformly most efficient diagnostic predictor. Notice that when the likelihoods of the case and control populations are confined to one value, the most efficient diagnostic predictor reads

$$\frac{L(x, \theta_0)}{\max(L(x, \theta_0), L(x, \theta_1))} < \lambda.$$

On the other hand, the most efficient diagnostic predictor according to Theorem 3.1 is

$$L(x, \theta_1) > c^* L(x, \theta_0).$$

Under this setting, we can clearly deduce that the two most efficient diagnostic predictors are identical, after a few steps of simple algebra derivation.

As shown in the previous examples, one discernible feature in the optimization process is the reduction of the data information from the n observations of the original data to one diagnostic predictor. In other words, the optimizing process with the evaluation standard becomes a process that reduces the dimension of the data toward a diagnostic predictor. Given this new perspective, when a sufficient statistic with lower dimension is available in the likelihood ratio expression, Theorem 3.1 can be simplified as follows to reduce the diagnostic predictor into a lower dimensional function.

Theorem 3.2 *Assume that the underlying model (pdf or pmf) of a set of data X is $f(x|\theta) \in \{f(x|case), f(x|healthy)\}$. Denote $T(X)$ a sufficient statistic for θ, and $g_i(t)$, $i = 0, 1$ the pmf (or pdf) of T corresponding to healthy and case populations, respectively. Then, the most efficient diagnostic measure becomes*

$$D(t) = \frac{g_1(t)}{g_0(t)},$$

with specificity

$$P(D > c|healthy) = 1 - \alpha.$$

Implementing the above theorem relies on the availability of a sufficient statistic that may reduce the dimension of the data while maintaining sufficient likelihood information. Since identifying a sufficient statistic is a key in the implementation of the above theorem, it is relevant to mention the factorization theorem, which involves the dimensional reduction process while preserving data sufficiency.

Theorem 3.3 Factorization theorem: *Let $f(x|\theta)$ denote the underpinning model (pmf or pdf) of a sample $\mathbf{x}$. A statistic $T(\mathbf{x})$ is sufficient for the unknown parameter θ if and only if it satisfies the following condition. There exist functions $g(t|\theta)$ and $h(\mathbf{x})$ such that, for all sample points and all permissible values of the parameter θ, the joint density can be decomposed into the product of information about the unknown parameter and information on the rest of the sample.*

$$f(\mathbf{x}|\theta) = g(T(\mathbf{x})|\theta)h(\mathbf{x}).$$

The proof and discussions on the Factorization Theorem can be found in Lehmann and Romano [81] or Casella and Berger [16]. Since this book focuses more on statistical prediction and machine learning, we elect not to pursue the theory of sufficient statistics in this book.

As pointed out in Example 3.4, we are often confronted with situations where the disease and healthy populations are referred to a range (instead of a value) of the observations. Under this scenario, the optimizing process discussed above cannot be directly applied. We shall now discuss the concept of the uniformly most efficient diagnostic predictor for a range of the patient healthy readings. With this objective in mind, we need the following concept.

Definition 3.4 Monotone Likelihood Ratio: *Assume that we have a set of data X that follows a family of underlying models (pmfs or pdfs) characterized by an unknown parameter $\theta \in \Theta$. Let T be a sufficient statistic of θ with the likelihood function $g(t|\theta) \in \{g(t|\theta) : \theta \in \Theta\}$. The monotone likelihood ratio property refers to the following property of the model of T: For any two points in the parameter space, $\theta_2 > \theta_1$, the likelihood ratio*

$$\lambda(t) = g(t|\theta_2)/g(t|\theta_1)$$

is a monotone function of t in the domain $\{t : g(t|\theta_1) > 0 \text{ or } g(t|\theta_2) > 0, \quad \theta_i \in \Theta\}$.

With the monotone likelihood ratio property, the process of optimizing sensitivity and specificity for a diagnostic predictor can be formulated as follows. A similar version to this result in hypothesis testing is the Carlin-Rubin theorem. For this reason, the following theorem is sometimes referred to as the *adapted Carlin-Rubin theorem* in sensitivity-specificity analysis.

Theorem 3.4 *Assume that the underlying model of the observations can be characterized by a density function $f(x|\theta)$ with a parameter θ. If $f(x|\theta)$ has the monotone (increasing) likelihood ratio property, the corresponding sensitivity is a nondecreasing function of the parameter (θ) with the setting*

$$\{D > c\} = \{T > t\},$$

and

$$Sensitivity(\theta_1) \geq Sensitivity(\theta_2),$$

when $\theta_1 > \theta_2$.

Proof: To show that the sensitivity is a non-decreasing function of the threshold, we consider the proof for continuous distributions for convenience. However, the proof is also applicable to discrete MLR families.

For $\theta_1 > \theta_2$, define

$$F(t|\theta, \ case) = 1 - P_\theta(T > t|\theta, \ case).$$

It suffices to show $F(t|\theta_1) \leq F(t|\theta_2)$, where F is the distribution function of T with parameter θ. Now

$$\frac{d}{dt}[F(t|\theta_1) - F(t|\theta_2)] = f(t|\theta_1) - f(t|\theta_2) = f(t|\theta_2)[\frac{f(t|\theta_1)}{f(t|\theta_2)} - 1].$$

Because $f(t|\theta)$ has MLR property, the ratio on the right-hand side is increasing with t, and the derivative can only change signs from negative to positive. This indicates that any interior extreme is a minimum, and the highest point of the function

$$g(t) = F(t|\theta_1) - F(t|\theta_2)$$

is located at $+\infty$ or $-\infty$.

$$g(-\infty) = F(-\infty|\theta_1) - F(-\infty|\theta_2) = 0,$$

$$g(\infty) = F(\infty|\theta_1) - F(\infty|\theta_2) = 0,$$

Thus, $g(t) \leq 0$ and

$$F(t|\theta_1) \leq F(t|\theta_2),$$

which is tantamount to

$$Sensitivity(\theta_1) = 1 - F(t|\theta_1) \geq 1 - F(t|\theta_2) = Sensitivity(\theta_2).$$

With Theorem 3.4, we have the following theorem (adapted Karlin-Rubin theorem) for the construction of the most efficient diagnostic predictor.

Theorem 3.5 *Assume that the underlying model of a set of data X is governed by a function characterized by patient feature $\theta \in R$. Consider a classification problem formulated as healthy $\theta \leq \theta_0$ and case $\theta > \theta_0$. Suppose that T is a sufficient statistic for θ, and the family of pmfs or pdfs of T has the MLR (Monotone Likelihood Ratio) property. If large value of D is associated with the disease, the most efficient diagnostic predictor is*

$$\{D > c\} = \{T > t_0\},$$

where D is the diagnostic measure and c is the diagnostic threshold. The value t_0 is determined according to the following condition,

$$Specificity = P(D \leq c | healthy) = P(T \leq t_0 | healthy) = 1 - \alpha.$$

Proof: Let $\beta(\theta) = P_\theta(T > t_0)$ be the sensitivity of the diagnostic predictor. Fix any value of the parameter $\theta' > \theta_0$, and consider a simple prediction problem on $\theta = \theta_0$ versus $\theta = \theta'$, since the underlying model (the family of pmfs or pdfs) of T is assumed to have the MLR property, by Theorem 3.4, $\beta(\theta)$ is a non-decreasing of θ, so we have

i) $\sup_{\theta \leq \theta_0} \beta(\theta) = \beta(\theta_0) = \alpha$, hence the specificity of the diagnostic predictor is $1 - \alpha$.

ii) If we define

$$k^* = \inf_{t \in \mathcal{T}} \frac{g(t | \theta')}{g(t | \theta_0)}$$

where $\mathcal{T} = \{t > t_0$ and either $g(t | \theta') > 0$ or $g(t | \theta_0) > 0\}$, it follows that

$$T > t_0 \Leftrightarrow \frac{g(t | \theta')}{g(t | \theta_0)} > k^*.$$

By Theorem 3.1, Parts (i) and (ii) imply that $\beta(\theta') > \beta^*(\theta')$, where $\beta^*(\theta)$ is the sensitivity of any other diagnostic predictor with specificity at $1 - \alpha$ level. Since θ' is arbitrary, the diagnostic predictor is the most efficient diagnostic predictor with specificity at the level $1 - \alpha$.

By an analogous argument, the following theorem can be derived.

Theorem 3.6 *Assume that the underlying model of a set of data X is governed by a function characterized by patient feature $\theta \in R$. Consider a classification problem formulated as healthy $\theta \leq \theta_0$ and case $\theta > \theta_0$. Suppose that T is a sufficient statistic for θ, and the family of pmfs or pdfs of T has the MLR (Monotone Likelihood Ratio) property. If small value of D is associated with the disease, the most efficient diagnostic predictor is*

$$\{D < c\} = \{T < t_0\},$$

where D is the diagnostic measure, c is the diagnostic threshold, and t_0 is the value that satisfies

$$Specificity = P(D \geq c | healthy) = P(T \geq t_0 | healthy) = 1 - \alpha.$$

The following example (a follow-up discussion on Example 3.4) demonstrates an application of the *adapted Carlin-Rubin theorem* in the optimization process toward the uniformly most efficient diagnostic predictor for a given specificity level $1 - \alpha$.

Example 3.5 *Let* $X_1......X_n$ *be a random sample of blood test LDL-cholesterol readings from a population. Assume that the readings follow a population* $N(\mu, 1)$ *with an unknown common mean LDL-Cholesterol level,* μ. *We are interested in showing that the solution in Example 3.4 is indeed the most efficient diagnostic predictor with specificity at 95% level.*

Solution: Since the normal model has the monotone likelihood ratio property, and high readings of LDL-Cholesterol level are associated with coronary heart disease, by the *adapted Carlin-Rubin theorem* (Theorem 3.4), the most efficient diagnostic predictor satisfies the condition

$$\{D > c\} = \{\frac{\overline{X} - 130}{\frac{1}{\sqrt{n}}} > 1.645\}.$$

This is equivalent to

$$\overline{X} > 130 + \frac{1.645}{\sqrt{n}},$$

the most efficient diagnostic measure is the sample mean LDL-Cholesterol level, and the diagnostic outcome is positive then the sample mean level reaches the corresponding threshold.

3.3   Two-ended diagnostic measures

The preceding section discusses methods to find the most efficient diagnostic predictor for one-ended extremes. When large values of the diagnostic measurement are associated with the disease such as LDL-Cholesterol level for coronary heart disease, it is the upper extreme. Alternatively, when low values of the predictor are associated with the disease, such as the RBC (red blood cell count) for leukemia, it is the lower extreme. However, in practice, there are many scenarios where both low and high measurements are associated with a disease. For instance, consider the reading of *bun to creatinine ratio* in a blood testing report. A high bun to creatinine ratio indicates conditions that lead to decreased blood flow to the kidney. On the other hand, a low bun to creatinine ratio implies increasing creatinine blood level, which also indicates kidney damage or kidney failure. In this case, we need to find a two-ended diagnostic predictor that can efficiently diagnose the disease. For example, assume that the healthy range for the *bun to creatinine ratio* is from

10:1 to 20:1, a blood test report with either too high (exceeding 20) or too low (below 10) *bun to creatinine ratio* is an indication of kidney failure.

Following the concept of optimization that we discussed in Section 3.1, for convenience, we continue with the idea of UMEP (uniformly most efficient predictor), and treat that as an example to illustrate the principle of restricted optimization when the underlying model is assumed.

As discussed in Theorem 3.4, the *adapted Karlin-Rubin* theorem ensures that the existence of an optimal solution for one-ended diagnostic measures when the underlying model has the MLR property. However, the story is different for the two-ended extreme problem. As shown in the following example, although the *adapted Karlin-Rubin* theorem is convenient in the derivation of optimal diagnostic predictor for one-ended extremes on the diagnostic measures, when we consider two-ended diagnostic predictors, the global optimal solution does not exist.

Example 3.6 *Let $X_1......X_n$ be a set of blood test readings on the bun to creatinine ratios. Assume that $X_i \sim N(\theta, \sigma^2)$ with known variance $\sigma^2 = 1$. Consider testing the prediction of healthy $\theta = \theta_0$ versus kidney disease $\theta \neq \theta_0$ for a given constant θ_0. Given a pre-fixed level of specificity $1 - \alpha$, we are interested in identifying the most efficient diagnostic predictor that satisfies*

$$P(\text{Claiming healthy}|\text{healthy}) \geq 1 - \alpha. \tag{3.4}$$

The optimal solution for this problem does not exist. To see this point, consider another parameter value $\theta_1 < \theta_0$ (for instance, two different values of *bun to creatinine ratios*), by the *adapted Karlin-Rubin* theorem, the uniformly most efficient predictor reads:
Claiming kidney diseases when

$$\bar{X} < -\sigma z_\alpha/\sqrt{n} + \theta_0.$$

This predictor has the highest sensitivity at the *bun to creatinine ratio* θ_1 among all predictors satisfying equation (3.4). We may call this Predictor-1. By the uniqueness of UME predictor, if a UMEP exists for this problem, it must almost surely be Predictor-1.

Now consider a different predictor, Predictor-2, which claims diseases when

$$\bar{X} > \sigma z_\alpha/\sqrt{n} + \theta_0.$$

Obviously, Predictor-2 also has the specificity level at $1 - \alpha$. We can now compare the sensitivity of the two predictors, Predictor-1 and Predictor-2 as follows.

Let $\beta_i(\theta)$ be the sensitivity function of predictor i with $i = 1, 2$. For any bun-to-creatinine-ratio $\theta_2 > \theta_0$, denote the sensitivity associated with

Predictor-i as $\beta_i(\theta)$, we have

$$\beta_2(\theta_2) = P_{\theta_2}(\bar{X} > \frac{z_\alpha \sigma}{\sqrt{n}} + \theta_0)$$

$$= P_{\theta_2}(\frac{\bar{X} - \theta_2}{\sigma/\sqrt{n}} > z_\alpha + \frac{\theta_0 - \theta_2}{\sigma/\sqrt{n}})$$

$$> P(Z > z_\alpha)$$

$$= P(Z < -z_\alpha)$$

$$> P_{\theta_2}(\frac{\bar{X} - \theta_2}{\sigma/\sqrt{n}} < -z_\alpha + \frac{\theta_0 - \theta_2}{\sigma/\sqrt{n}})$$

$$= P_{\theta_2}(\bar{X} < -\frac{\sigma z_\alpha}{\sqrt{n}} + \theta_0)$$

$$= \beta_1(\theta_2)$$

This shows that the sensitivity of Predictor-1 at the bun-to-creatinine-ratio, θ_2, is lower than its counterpart. Thus, Predictor-1 is not a UMEP with specificity $1-\alpha$. This is in contradiction with the earlier statement that Predictor-1 is the UMEP. Therefore, the UMEP with specificity level $1 - \alpha$ does not exist for the two-ended diagnostic scenarios.

When the UMEP does not exist among all the predictors that have specificity $1 - \alpha$, instead of seeking the global optimal predictor, we may put a restriction on the domain of predictors in the optimization process, and seek for a local optimal solution within a confined domain of diagnostic measures. For example, we may consider the search of optimal solution to a subgroup of predictors named *decent predictor* defined below.

Definition 3.5 Decent predictor: *When predicting disease populations with a diagnostic measure, a decent predictor is a predictor satisfying the condition that the probability of correctly diagnosing a sick patient is higher than the probability of incorrectly diagnosing a healthy subject.*

The idea of decent predictor is similar to the concept of unbiased test defined for the content of hypothesis testing. A formal mathematical definition of the unbiased test can be found in [81].

The Definition 3.5 for a decent predictor can be expressed as

$$Sensitivity(\phi, \theta) > \alpha, \quad \text{and} \quad P(false\ positive) < \alpha,$$

or

$$Sensitivity(\phi, \theta) > \alpha, \quad \text{and} \quad Specificity \geq 1 - \alpha.$$

As usual, $\phi(\mathbf{X})$ is the prediction function. When the diagnosed outcome is positive, $\phi(\mathbf{X}) = 1$, otherwise $\phi(\mathbf{X}) = 0$.

From the Definition 3.5, any UMEP with specificity $1 - \alpha$ is an UMEDP (uniformly most efficient and decent predictor). On the other hand, with additional restriction of being a decent predictor, for prediction problems where

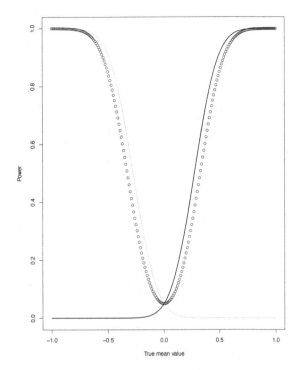

FIGURE 3.3
Sensitivity functions $\beta_1(\theta)$, $\beta_2(\theta)$, and $\beta_3(\theta)$.

UMEP does not exist, UMEDP may exist. We illustrate this point with the following example.

As shown in Definition 3.5, the concept of *decent predictor* is a natural requirement for a diagnostic predictor. Namely, the probability of correctly detecting the diseases subject should be at least as large as the probability of incorrectly diagnose a healthy subject. With such a restriction, we are able to search for a UMEP with specificity $1 - \alpha$ among the class of *decent predictors*. Such a restricted optimization procedure results in the construction of the following UMEDP.

Consider the sensitivity function $\beta(\theta)$ of a diagnostic predictor Predictor-3 which diagnoses a subject as having the disease when

$$\frac{|\bar{X} - \theta_0|}{\sigma/\sqrt{n}} > z_{\alpha/2}.$$

Figure 3.3 shows the three sensitivity curves corresponding to the three predictors discussed above. The dotted curve is for $\beta_3(\theta)$, the black curve is for $\beta_2(\theta)$, and $\beta_1(\theta)$ has the gray curve. As shown in the diagram, the dotted curve, although not as powerful as the gray or the black curve at some points, is able to achieve its own local optimal sensitivity when the diagnostic measure gets either larger or smaller. However, for the other two curves, the sensitivity drops below the probability of false positive rate α when the diagnostic measure gets larger for $\beta_1(\theta)$ (or smaller for $\beta_2(\theta)$). Thus, they are not *decent predictors*.

We will prove that the dotted sensitivity curve is indeed the restricted optimal curve in Example 3.7 after a discussion on the following theorem, which sets the connection between the global optimal solution (a UMEP level $1 - \alpha$ predictor) and a restricted optimal solution. The latter is the UMEDP level $1 - \alpha$ predictor for this type of diagnostic measures. Theorem 3.7 shows a way to identifying a UMEDP by considering UMEP in the boundary between the diseased and the healthy populations.

Theorem 3.7 *For a model characterized by a parameter θ, assume that the corresponding sensitivity function of every predictor is continuous. For level $1 - \alpha$ diagnostic predictors on the diseased and healthy populations, a predictor ϕ is a UMEDP (uniformly most efficient and decent predictor) if it is a UMEP among all predictors satisfying the following equation (3.5) on the diagnostic boundary,*

$$\beta_\phi(\theta) = \alpha, \tag{3.5}$$

where $\theta \in \omega$, and ω is the set of the diagnostic boundary between diseased and healthy subjects.

For example, when the range of healthy *bun to creatinine ratio* is from 10:1 to 20:1, the diagnostic boundary is formed by two values $\{10, 20\}$.

Proof *The class of predictors satisfying (3.5) contains the set of decent predictors, hence the sensitivity of the UMEP ϕ_0 is at least as high as the sensitivity of any decent predictor with the same specificity. On the other hand, the UMEP predictor ϕ_0 is decent itself. This is because it is uniformly at least as sensitive as the predictor $\phi(x) \equiv \alpha$, where $\phi(x)$ is a special predictor that claims a disease case by flipping a biased coin. When the outcome is a head after flipping the coin with*

$$P(Head) = \alpha,$$

$\phi(x)$ *claims diseases.*

When additional information is available, we generally expect to have a more accurate prediction for the optimizing problem of interest. For instance, when the underlying distribution of diagnostic measures can be assumed to be in the exponential family, the restricted optimal solution discussed in the illustrating UMEDP example can be explicitly constructed according to the following theorem.

Theorem 3.8 Restricted optimization for UMEDP *Assume that the underlying model for a set of data* $\mathbf{X}$ *can be expressed explicitly as an exponential family, with the joint density characterized by a parameter θ:*

$$f(x) = c(\theta)e^{\theta T(x)}h(x) \quad c(\theta) > 0, \tag{3.6}$$

for the prediction problem with two-ended diagnostic measures. If there exist two constants λ_1 and λ_2 such that

$$P(\mathbf{X} \in \{T > \lambda_1\} \bigcup \{T < \lambda_2\}|healthy) = 1 - \alpha,$$

then the UMEDP is the diagnostic predictor shaped by the region

$$D = \{T > \lambda_1\} \bigcup \{T < \lambda_2\}.$$

Proof *Consider the area $A = \{f(x) > \lambda g(x)\}$, where $f(x)$ is the likelihood function for the diseased population, and $g(x)$ is the one for the healthy population. For any predictor with specificity $1 - \alpha$, we have*

$$(I_B - I_A)(f(x) - \lambda g(x)) \leq 0,$$

where I_A is the indicator function of set A.

Now. since the underlying model is assumed to be in the form of (3.6), we have

$$\int (I_B - I_A)(f(x) - \lambda g(x))dx \leq 0,$$

implies

$$\int I_B f(x)dx - \int I_A f(x)dx \leq \lambda(\int I_B g(x)dx - \int I_A g(x)dx) \leq 0,$$

which means that the predictor corresponding to the indicator function of the region A is a uniformly most sensitive predictor.

Now

$$\frac{f(x)}{g(x)} = \frac{c(\theta)e^{\theta Th(x)}}{c(\theta_0)e^{\theta_0 Th(x)}} = d(\theta, \theta_0)e^{(\theta - \theta_0)T}$$

Notice that

$$Set A \iff \{x : (\theta - \theta_0)T(x) > c^{**}\}$$
$$\iff \{T(x) > \lambda^{***}\} \bigcup \{T(x) > \lambda^{**}\}$$

we have

$$P(\{T(x) > \lambda^{***}\} \bigcup \{T(x) > \lambda^{**}\}|\theta_0) = \alpha.$$

In conjunction with Theorem 3.7, the condition on (3.5) is satisfied, thus, the diagnostic predictor $\{T(x) > \lambda^{***}\} \bigcup \{T(x) > \lambda^{**}\}$ is UMEDP.

The above theorem identifies the general shape of a restricted optimization problem for the uniformly most efficient decent predictor. However, even in this particular example, the determination of the explicit diagnostic predictor is still unspecified. In this context, we have the following theorem that provides a specific solution to construct the restricted optimizing diagnostic predictor, UMEDP.

Theorem 3.9 *Assume that the underlying model of a set of data* $\mathbf{X}$ *can be expressed explicitly as an exponential family with the joint likelihood characterized by a parameter* θ *as in (3.6),*

$$f(\mathbf{X}) = c(\theta)e^{\theta T(\mathbf{X})}h(\mathbf{X}) \quad c(\theta) > 0.$$

For the two-ended prediction/classification problem on diseased and healthy populations, the restricted optimizing solution (UMEDP) can be determined via the following two conditions:

$$E_{\theta_0}(\phi(\mathbf{X})) = \alpha \tag{3.7}$$

$$E_{\theta_0}(T(\mathbf{X})\phi(\mathbf{X})) = \alpha E_{\theta_0}(T(\mathbf{X})) \tag{3.8}$$

Proof: The sensitivity function is

$$\beta(\theta) = \int_{R^n} \phi(\mathbf{X})c(\theta)e^{\theta T(\mathbf{X})}h(\mathbf{X})d\mathbf{X}$$

Taking the derivative with respect to θ and evaluating at θ_0, yields,

$$\beta'(\theta_0) = \int_{R^n} \phi(\mathbf{X})c'(\theta_0)e^{\theta_0 T(\mathbf{X})}h(\mathbf{X})d\mathbf{X} + \int_{R^n} \phi(\mathbf{X})c(\theta_0)T(\mathbf{X})e^{\theta_0 T(\mathbf{X})}h(\mathbf{X})d\mathbf{X}$$

$$= \frac{c'(\theta_0)}{c(\theta_0)}E_{\theta_0}(\phi(\mathbf{X})) + E_{\theta_0}(\phi(\mathbf{X})T(\mathbf{X}))$$

$$= 0$$

The derivative is 0 at θ_0 because the sensitivity of any decent predictor is minimized at θ_0 (the diagnostic measure for the healthy population) by definition.

Since the above equation is true for any decent predictor, we consider

$$\phi(\mathbf{X}) \equiv \alpha,$$

this prediction function also defines a decent predictor: claiming diseased with probability α regardless of the sample. Now, we have:

$$E_{\theta_0}(T(\mathbf{X})) = -\frac{c'(\theta_0)}{c(\theta_0)}$$

Also, the first equality is implied by the condition of decent predictors, we have

$$E_{\theta_0}(\phi(\mathbf{X})T(\mathbf{X})) - \alpha E_{\theta_0}(T(\mathbf{X})) = 0.$$

This verifies the second equality in the theorem.

When the underlying model of the data set is a symmetric distribution, such as the normal model or Laplance model, the construction of the UMEDP restricted optimal solution can be simplified in the following theorem.

Theorem 3.10 *Following conditions of Theorem 3.9, when the underlying model of the data is symmetric about a constant r, the restricted optimal solution for the two-ended diagnostic prediction problem can be constructed by setting $\lambda_1 = 2r - \lambda_2$, where the values λ_1 and λ_2 are as defined in Theorem 3.8.*

Proof: Denote $g(t|\theta)$ the underlying model of the data, by (3.8), we have

$$E_{\theta_0}(T\phi) = \alpha E_{\theta_0}(T) \qquad (3.9)$$

Since $g(t|\theta)$ is symmetric about r, we have $E_{\theta_0}(T) = r$, so (3.9) becomes $E_{\theta_0}(T\phi) = \alpha r$, $\alpha = E_{\theta_0}(\phi)$, which is

$$E_{\theta_0}((T - r)\phi) = 0.$$

By Theorem 3.9, the UMEDP takes the form of claiming diseases when the diagnostic measure T is either too large or too small, denote

$$\phi = \begin{cases} 1, & T > \lambda_1 \text{ or } T < \lambda_2 \\ 0, & \text{otherwise.} \end{cases}$$

We have

$$\int_{-\infty}^{\lambda_2} (t - r)g(t|\theta_0)dt + \int_{\lambda}^{+\infty} (t - r)g(t|\theta_0)dt = 0.$$

By symmetry $g(t|\theta_0) = g(2r - t|\theta_0)$, we have

$$\int_{-\infty}^{\lambda_2} (t - r)g(t|\theta_0)dt - \int_{-\infty}^{2r-\lambda_1} (y - r)g(y|\theta_0)dy = 0.$$

Since $y - r = 2r - t - r = r - t$, $dy = -dt$.

Therefore

$$\int_{-\infty}^{\lambda_2} (t - r)g(t|\theta_0)dt = \int_{-\infty}^{2r-\lambda_1} (y - r)g(y|\theta_0)dy,$$

and $\lambda_2 = 2r - \lambda_1$. This concludes the proof of the theorem.

In what follows in this section, we shall provide two examples to illustrate how to perform the restricted optimization to construct the efficient and decent predictor/classifier, when the underlying model can be plausibly assumed.

Example 3.7 *Assume that the distribution of the bun-to-creatinine-ratio follow a normal model with a given standard deviation σ and healthy mean ratio μ_0. In this case, with the notation defined in Theorem 3.10, $\lambda_1 = z_{\alpha/2}$.*

Solution: Notice that

$$E_{\mu_0}(T) = E_{\mu_0}(\frac{\bar{X} - \mu_0}{\sigma/\sqrt{n}}) = 0,$$

which implies that

$$\lambda_2 = -\lambda_1.$$

We also have

$$P(Z > \lambda_1) + P(Z < \lambda_2)$$
$$= \alpha$$
$$= P(Z > \lambda_1) + P(Z < -\lambda_1)$$
$$= 2P(Z > \lambda_1),$$

so

$$\lambda_1 = z_{\alpha/2}.$$

This example offers a theoretical justification to the existence of UMEDP (the dotted curve) in Figure 3.3. Notice that the functioning of Example 3.7 is grounded on the assumption that the stability of the diagnostic measure (σ) is a given value. In practice, the measurement stability score is an unknown value. As a follow-up discussion, we shall describe a method pertaining to the prediction of measurement stability scores. In particular, we discuss an example predicting the risk level of an investment portfolio.

Example 3.8 *Let* $\mathbf{X} = (X_1......X_n)$ *be a random sample of stock returns of an investment portfolio. For convenience, assume that in a short period of time with market equilibrium, the data follow a normal population with mean zero and unknown risk index* σ^2, *where* σ^2 *reflects the measurement stability of the portfolio. We are interested in finding the restricted optimal solution in the setting for UMEDP in predicting* $\sigma = \sigma_0$ *(the investment risk is at an envisaged level) versus* $\sigma \neq \sigma_0$ *(the envisaged level* σ_0 *is either under-predicted or over-predicted the true but unknown investment risk* σ).

The requirement is that the predictor is a decent predictor with specificity at level $1 - \alpha$.

Solution: The likelihood

$$L = (\frac{1}{\sqrt{2\pi}\sigma})^n e^{-\frac{\sum_{i=1}^n x_i^2}{2\sigma^2}}$$

belongs to the exponential family, so $T(\mathbf{X}) = \sum_{i=1}^n X_i^2$. By Theorem 3.8, the decent predictor is given by

$$\phi = \begin{cases} 1, & T > \lambda_1 \text{ or } T < \lambda_2 \\ 0, & \text{otherwise.} \end{cases}$$

If the investment risk is really at σ_0 level,

$$P_{\sigma_0}(T < \lambda_2 \text{ or } T > \lambda_1) = \alpha,$$

and

$$P_{\sigma_0}(\lambda_2 < T < \lambda_1) = 1 - \alpha.$$

Now, let $Y = \frac{T}{\sigma_0^2}$, the distribution of Y follows the χ_n^2 under the assumption that the investment risk is σ_0.

$$\phi = \begin{cases} 1, & Y > d_2 \text{ or } Y < d_1 \\ 0, & \text{otherwise.} \end{cases}$$

By Theorem 3.9,

$$\begin{aligned} E_{\sigma_0}(\phi Y) &= \int_{\{y<d_1\}\cup\{y>d_2\}} y f_Y(y) dy \\ &= \alpha E_{\sigma_0}(Y) \\ &= n\alpha, \end{aligned}$$

and

$$E_{\sigma_0}(\phi) = \alpha.$$

Thus, the two conditions for UMEDP are:

$$\int_{d_1}^{d_2} \frac{1}{2^{\frac{n}{2}}\Gamma(\frac{n}{2})} y^{\frac{n}{2}} e^{-\frac{y}{2}} dy = n(1-\alpha) \tag{3.10}$$

$$\int_{d_1}^{d_2} \frac{1}{2^{\frac{n}{2}}\Gamma(\frac{n}{2})} y^{\frac{n}{2}} e^{-\frac{y}{2}-1} dy = 1-\alpha \tag{3.11}$$

By (3.10),

$$\int_{d_1}^{d_2} \frac{-2}{2^{\frac{n}{2}}\Gamma(\frac{n}{2})} y^{\frac{n}{2}} d e^{-\frac{y}{2}} = n(1-\alpha),$$

which is equivalent to

$$\frac{-2}{2^{\frac{n}{2}}\Gamma(\frac{n}{2})} y^{\frac{n}{2}} e^{-\frac{y}{2}}\Big|_{d_1}^{d_2} - \int_{d_1}^{d_2} \frac{-2}{2^{\frac{n}{2}}\Gamma(\frac{n}{2})} (\frac{n}{2}) y^{\frac{n}{2}-1} e^{-\frac{y}{2}} dy = n(1-\alpha).$$

By (3.11),

$$\frac{-2}{2^{\frac{n}{2}}\Gamma(\frac{n}{2})} y^{\frac{n}{2}} e^{-\frac{y}{2}}\Big|_{d_1}^{d_2} + n(1-\alpha) = n(1-\alpha).$$

So d_1 and d_2 satisfy the following equation,

$$\frac{(-2)d_2^{-\frac{n}{2}} e^{-\frac{d_2}{2}}}{2^{\frac{n}{2}}\Gamma(\frac{n}{2})} = \frac{(-2)d_1^{-\frac{n}{2}} e^{-\frac{d_1}{2}}}{2^{\frac{n}{2}}\Gamma(\frac{n}{2})} \Rightarrow d_1^{-\frac{n}{2}} e^{-\frac{d_1}{2}} = d_2^{-\frac{n}{2}} e^{-\frac{d_2}{2}}.$$

Therefore, the UME decent predictor is

$$\{\mathbf{X} : \frac{\sum X_i^2}{\sigma_0^2} < d_1\} \bigcup \{\mathbf{X} : \frac{\sum X_i^2}{\sigma_0^2} < d_1\},$$

where the cutoff values d_1 and d_2 satisfy

$$P(\chi_n^2 < d_1) + P(\chi_n^2 > d_2) = \alpha,$$

and

$$d_1^{-\frac{n}{2}} e^{-\frac{d_1}{2}} = d_2^{-\frac{n}{2}} e^{-\frac{d_2}{2}}.$$

This section focuses on the method of restricted optimization (the most efficient and decent diagnostic predictor) when the global optimal solution does not exist. It shows that by adding an additional condition (a decent predictor in the way that the probability of correct diagnosis exceeds the probability of false positive), we can restrict the optimization domain on predictors satisfying certain conditions (controlling the specificity at level $1 - \alpha$), and find a restricted optimal solution (uniformly most efficient decent diagnostic predictor, UMEDP). The next section will follow up with optimization in the case where nuisance condition exists.

3.4   UMEDP with confounding factors

When the global optimization does not exist due to the condition that the diagnostic measure has two-ended extremes related to the diseases, we use restricted optimization (such as the uniformly most efficient and decent predictors, UMEDP). Yet, the two-ended diagnostic measure is not the only cause for the nonexistence of UMEP (in the sense of achieving the highest sensitivity for a given specificity). In this section, we shall discuss a scenario in which nuisance parameters or confounding factors also result in nonexistence of UMEP. The concept of restricted optimization such as the decent predictor defined in Definition 3.5 will be applied again to seek for a local optimal solution.

In model-based analytics, when an assumed model involves various parameters, some of the parameters are of primary interest for data analysis, while others are not of immediate interest (but still play critical roles in the assumed model). For instance, in the optimization process with restrictions for UMEDP, when predicting the mean vector, elements in the covariance matrix (which measure the variation and inter-correlations among the variables) are regarded as nuisance parameters. The restricted optimization process involving nuisance parameters is another critical issue in statistical prediction.

We start with a simple example of the inference about the population mean while the population variation can not be plausibly assumed. The population variation serves as a nuisance parameter for the prediction problem in this case. The example shows how a confounding nuisance factor alters the existence of the global optimal solution (UMEP).

Example 3.9 *Consider the uniformly most efficient diagnostic predictor in Example 3.5, where we assume that the variation of the LDL-Cholesterol reading is 1. In fact, the variation of the LDL-Cholesterol reading is unknown in practice. As discussed in Example 3.5, for a given variation index σ_0, the UMEP with specificity $1 - \alpha$ reads*

$$\{\mathbf{X} : \frac{\bar{X} - 130}{\sigma_0/\sqrt{n}} > Z_\alpha\}.$$

However, the actual standard deviation may not be σ_0.

If the true but unknown standard deviation doubles the assumed value, say, $\sigma = 2\sigma_0$, the probability of false positive rate of the UMEP with specificity $1-\alpha$ test becomes

$$P_{\mu_0}(\{\mathbf{X} : \frac{\bar{X} - \mu_0}{\sigma_0/\sqrt{n}} > Z_\alpha\})$$

$$=P_{\mu_0}(\frac{\bar{X} - \mu_0}{\sigma_0/\sqrt{n}} > Z_\alpha)$$

$$=P_{\mu_0}(\frac{\bar{X} - \mu_0}{2\sigma_0/\sqrt{n}} > \frac{Z_\alpha}{2})$$

$$=P(Z > \frac{Z_\alpha}{2})$$

$$>\alpha.$$

In this case, when the probability of false positive rate is larger than α, the corresponding specificity becomes less than $1-\alpha$, hence the UMEP with incorrectly assumed standard deviation is not a diagnostic prediction at the nominal specificity level $1 - \alpha$ for the prediction problem.

This example shows that the UMEP does not exist in the process of predicting μ, when σ is unknown. In this case, the unknown standard deviation σ serves as a nuisance parameter. We shall now introduce a theorem that can be viewed as an example of restricted optimization under the presence of nuisance parameters. The method is similar to the discussion on the example of two-ended diagnostic measures in the preceding section.

Theorem 3.11 *Assume that the underlying model of the data set X follows an exponential family*

$$f(X) = h(x)exp\{\theta U(x) + \sum_{i=1}^{k} v_i T_i(x) + c(\theta, \mathbf{v})\},$$

where θ is the parameter of interest, and $\mathbf{v}$ is a vector of nuisance parameters. Consider a diagnostic prediction problem for disease subjects classified as healthy $\theta < \theta_0$, and diseased $\theta > \theta_0$. Further, assume that $W(U, \mathbf{T})$ is a monotone function in U for each vector of statistics $\mathbf{T}$, $\mathbf{T} = (T_1, ..., T_k)$. If $W(U, \mathbf{T})$ and $\mathbf{T}$ are independent on the boundary θ_0, then the diagnostic predictor

$$\phi(W) = \begin{cases} 1, & W > C_\alpha \\ \gamma_\alpha, & W = C_\alpha \\ 0 & W < C_\alpha \end{cases}$$

is the uniformly most efficient decent predictor, where the constants C_α and γ_α are determined such that the specificity of $\phi(W)$ is $1 - \alpha$.

Proof: For any fixed $\mathbf{T} = (T_1(x), ..., T_k(x))$, $P_\theta(x)$ has the monotone likelihood ration property in terms of U, by the *adapted Karlin-Rubin* theorem, the UMEP reads

$$\phi^*(U|\mathbf{T}) = \begin{cases} 1, & U > \xi_\alpha(\mathbf{T}) \\ \gamma_\alpha(\mathbf{T}), & U = \xi_\alpha(\mathbf{T}) \\ 0 & U < \xi_\alpha(\mathbf{T}) \end{cases}$$

where $\xi_\alpha(\mathbf{T})$ and $\gamma_\alpha(\mathbf{T})$ are determined by

$$E_{\theta_0, \mathbf{v}}(\phi^*(U|\mathbf{T})|\mathbf{T} = t) = \alpha$$

for all $\mathbf{T} = (t_1, ..., t_k)$. Now, if W is a monotone function (without loss of generality, assume that it is an increasing function) of U,

$$\phi^{**}(U|\mathbf{T}) = \begin{cases} 1, & W > \xi'_\alpha(\mathbf{T}) \\ \gamma'_\alpha(\mathbf{T}), & W = \xi'_\alpha(\mathbf{T}) \\ 0 & W < \xi'_\alpha(\mathbf{T}) \end{cases}$$

is a UMEDP.

Notice that

$$\phi^{**}(U(\mathbf{T})|\mathbf{T}) = \begin{cases} 1, & W > \xi'_\alpha(\mathbf{T}) \\ \gamma'_\alpha(\mathbf{T}), & W = \xi'_\alpha(\mathbf{T}) \\ 0 & W < \xi'_\alpha(\mathbf{T}) \end{cases}$$

satisfies

$$E_{\theta_0}(\phi^{**}|\mathbf{T}) = P_{\theta_0}(W > \xi'_\alpha(\mathbf{T})|\mathbf{T}) + \gamma'_\alpha(\mathbf{T})P_{\theta_0}(W = \xi'_\alpha(\mathbf{T})|\mathbf{T}).$$

Now, since variables W and $\mathbf{T}$ are independent at the boundary θ_0, there exist constants γ_α and C_α that are not functions of $\mathbf{T}$, so that

$$E_{\theta_0}(\phi^{**}(W)) = P_{\theta_0}(W(U, \mathbf{T}) > C_\alpha) + \gamma_\alpha P(W = C_\alpha) = \alpha.$$

Thus, the test

$$\phi^{***}(W) = \begin{cases} 1, & W > C_\alpha \\ \gamma_\alpha, & W = C_\alpha \\ 0 & W < C_\alpha \end{cases}$$

is the UMEP of diagnostic predictors satisfying

$$E_{\theta_0}(\phi^{***}) = \alpha$$

for any vector of nuisance components $\mathbf{v}$. This shows that the diagnostic predictor ϕ^{***} is an optimal solution restricted to the set of *decent predictors*.

We can now use Theorem 3.11 to find the uniformly most efficient decent diagnostic predictor for the LDL-Cholesterol prediction in Example 3.5, when the standard deviation is allowed to change within its permissible domain.

Example 3.10 Restricted optimization with nuisance parameters.
Consider a set of observations $X_1, \ldots\ldots X_n \sim N(\mu, \sigma^2)$ where σ is unknown. Assume that the cutoff threshold for the diagnostic predictor is μ_0 (such as the 130 mg/dL for the LDL-cholesterol readings). The restricted optimal solution in the prediction of diseased population on the basis of diagnostic measures becomes

$$\phi = \begin{cases} 1, & T > t_{n-1}(\alpha) \\ 0 & otherwise \end{cases}$$

where

$$T = \frac{\sqrt{n}(\bar{X} - \mu_0)}{s},$$

$\bar{X}$ *is the sample mean reading and s is the sample standard deviation of the LDL-cholesterol readings.*

Solution: By Theorem 3.11, we need to rewrite the likelihood function as follows.

$$P_\theta(\mathbf{X}) = \left(\frac{1}{\sqrt{2\pi}\sigma}\right)^n e^{-\frac{1}{2\sigma^2}\left(\sum_{i=1}^n X_i^2 - 2\mu \sum_{i=1}^n X_i + n\mu^2\right)}.$$

Let $U = \bar{X} - \mu_0$, $T = \sum_{i=1}^n (X_i - \mu_0)^2$, for $\mu = \mu_0$, T is minimal sufficient for σ^2.

Consider

$$W = \frac{\sqrt{n}U}{\left[\frac{(T - nU^2)}{n-1}\right]^{\frac{1}{2}}}$$

at $\mu = \mu_0$, $W \sim t_{n-1}$. And W is ancillary of σ^2. Also, $T \sim \sigma^2 \chi_n^2$ at $\mu = \mu_0$, by Basu's theorem, W and T are independent at $\mu = \mu_0$.

Now, notice that

$$T - nU^2$$

$$= \sum_{i=1}^{n}(X_i - \mu_0)^2 + n(\bar{X} - \mu_0)^2$$

$$= \sum_{i=1}^{n}(X_i - \bar{X})^2$$

$$= (n-1)s^2 > 0,$$

we can decompose W in the following way when $U > 0$,

$$\log(W) = \log\sqrt{n} + \log(U) - \frac{1}{2}\left(\log\frac{1}{n-1} + \log(T - nU^2)\right).$$

Furthermore,

$$\frac{\partial\log(W)}{\partial U} = \frac{1}{U} - \frac{1}{2} \times \frac{(-2nU)}{T - nU^2}$$

$$= \frac{1}{U} + \frac{nU}{T - nU^2}$$

$$> 0,$$

so W increases as U increases when $U > 0$ for each T. When $U < 0$, letting $V = -U$ and applying the above arguments yields the conclusion that W is a monotone function of U in this example.

By Theorem 3.11, ϕ is the restricted optimization solution (UMEDP) for the prediction problem specified in the example.

According to the above example, a diagnostic predictor based on the usual Student t statistic is the best diagnostic predictor in the sense that it controls the specificity at $1 - \alpha$ level, maximizes the sensitivity, and satisfies the condition of *decent predictors*. In the context of hypothesis testing, this is parallel to the uniformly most powerful unbiased test in testing the mean with standard deviation as the nuisance parameter.

In the next section, we shall discuss another fundamental principle in restricted optimization, the invariant principle. It keeps the predicted result consistent and invariant when the same experimental subject is measured by different scales of the diagnostic measurement.

3.5 Efficient and invariant diagnostic predictors

The source of machine learning is data, and data comes from the observations and measurements of experimental subjects that we are interested in. With the same object, different measurement units (or scales) often make the data look different. One frequently asked question is the consistency of the prediction results when different scales or units are used in measuring the subjects in an experiment. It is necessary to have a prediction method that remains consistent for various measurements on the same object. For example, consider the comparison of body heights between college students and elementary school students. If a prediction method claims significant mean difference with heights measured in *cm*, we would expect a similar claim when the same objects are measured by *m*, because measuring with the scale of *cm* or *m* should not alter the fact that on average, college students are taller than elementary school students.

3.5.1 Invariant principle in data transformation

There is an excellent resource discussing this topic in the literature. We start with a few basic concepts on model invariant as defined in [81].

Definition 3.6 Model invariant: *Let g be a 1-to-1 transformation from the sample space to itself. For a set of data X, if the transformed data g(X) follows the same model as the model of the original data, the underlying model is called model invariant under the data transformation g.*

A mathematical definition according to [81] can be formulated as follows. Let X be a random variable taking values in a sample space according to a probability model from the family $P = \{P_\theta, \theta \in \Omega\}$. For a one-to-one transformation g from the sample space into itself. If, for each θ, the distribution of $X' = g(X)$ is a member of P, say $P_{\theta'}$, and $\bar{g}(\Omega) = \Omega$ (as θ travels through Ω, so does θ'), then the probability model P is invariant under transformation g.

It is related at this point to consider a class of permissible transformations (a special case of data transformation) defined below. This is because we are considering restricted optimization in this chapter, to achieve the local optimal solution for UMEP, we shall define the domain of restriction on the set of transformations as below.

Definition 3.7 Group: *A set of elements is called a* group *if it satisfies the following four conditions:*

1 *There is an operation defined for elements in $\mathcal{G}$, group multiplication. Namely, for any two elements $a, b \in \mathcal{G}$, there exists an element $c \in \mathcal{G}$ such that $c = ab$, where the element c is called the product of a and b and denoted as ab.*

2 *Group multiplication obeys the associative law. Namely for any three elements in $\mathcal{G}$, $(ab)c = a(bc)$.*

3 *There is an element in $\mathcal{G}$ called* identity, *such that $ae = ea$ for all $a \in \mathcal{G}$.*

4 *For each element a in G, there exist an* $a^{-1} \in \mathcal{G}$ *(its* inverse *in G), such that* $aa^{-1} = a^{-1}a = e$.

In the definition above, both the inverse a^{-1} of any element $a \in \mathcal{G}$ and the identity element $e \in \mathcal{G}$ can be shown to be unique.

When we consider transformations on a data set, it is helpful to utilize the concept on *transformation group*.

Definition 3.8 Transformation group: *A set of transformations* $\{g : g \in \mathcal{G}\}$ *from the sample space S to S is called a group of transformation in S if it satisfies the following conditions.*

1 *Inverse transformation is self-contained in the set, namely for every* $g \in \mathcal{G}$, *there is a* $g' \in \mathcal{G}$ *such that* $g'(g(x)) = x$ *for all* $x \in S$.

2 *Composition is self-contained in the set, namely for every* g *and* $g' \in \mathcal{G}$, *there exist a* $g'' \in \mathcal{G}$ *such that* $g'(g(x)) = g''(x)$.

3 *Identity transformation is self-contained in the set, namely* $e(x)$ *defined by* $e(x) = x$ *is in set* $\mathcal{G}$.

Example 3.11 *A class G of transformations is a* transformation group *if it is closed under both composition and inversion.*

Solution: It is straightforward to verify that transformation, is in fact, a group. In particular, note that the *identity transformation* $x \equiv x$ is a member of any transformation group G since $g \in G$ implies $g^{-1} \in G$ and hence $gg^{-1} \in G$, and by definition gg^{-1} is the identity. Note also that the inverse $(g^{-1})^{-1}$ of g^{-1} is g, so that gg^{-1} is also the identity.

Example 3.12 *The following are two groups of data transformations for two common distribution families.*

1 $G = \{X, n - X\}$ *for data following the binomial family* $Bin(n, p)$.

2 $G = \{X - a, a \in R\}$ *for data following the normal distribution family,* $N(\mu, \sigma^2)$.

Following the concept of transformation group, we can now extend the concept of model invariant on one transformation into model invariant for a group of transformations.

Definition 3.9 Group invariant: *If the underlying model of a set of data is invariant for every element of a class of transformations C, then the model is invariant under this set of data transformations, C.*

We use the following example to clarify the above concepts in a heuristic way for data transformations. Further details with advanced mathematical treatments can be found in [81].

One of the interesting applications of the definition above is the property that the Student-t test is invariant for any location and scale transformation. For example, if the sample mean is 1.72m with sample standard deviation of 0.5m for a random sample of 25 college male students, and correspondingly 1.64m (sample mean) with 0.4m (sample standard deviation) for 25 female students. Assuming that the population variations are the same for the two populations, the difference t-score is

$$\frac{1.72 - 1.64}{\sqrt{\frac{24*0.5^2+24*0.4^2}{48}}},$$

which is the same as the measurements that use cm and boot every subject by 10 cm:

$$\frac{172 - 164}{\sqrt{\frac{24*(50)^2+24*(40)^2}{48}}}.$$

In fact the above numerical example is a special case of the following invariant principle.

Example 3.13 Invariant principle: *The Student-t based diagnostic predictor is invariant under location and scale transformations.*

Solution: Let $\mathbf{X} = (X_1, ..., X_n)$ be the data set for analysis, the Student-t based diagnostic predictor is

$$t(\mathbf{X}) = \frac{\bar{X} - E(X_1)}{s_X/\sqrt{n}}.$$

Now the location and scale transformation data become

$$\mathbf{Y} = a\mathbf{1} + b\mathbf{X},$$

with the expected value $E(\mathbf{Y}) = a\mathbf{1} + bE(\mathbf{X})$. The variance of the transformed data reads,

$$
\begin{aligned}
s_Y^2 &= \frac{1}{n-1}\sum_{i=1}^{n}(Y_i - \bar{Y})^2 \\
&= \frac{1}{n-1}\sum_{i=1}^{n}(a + bX_i - a - b\bar{X})^2 \\
&= \frac{b^2}{n-1}\sum_{i=1}^{n}(X_i - \bar{X})^2 \\
&= b^2 s_X^2
\end{aligned}
$$

so $s_Y = bs_X$. Even with location transformation, the sample standard deviation of the transformed data is the scale transformation of the original sample standard deviation. Therefore the corresponding t-score becomes

$$t(\mathbf{Y}) = \frac{a + b\bar{X} - a - bE(X_1)}{bs_X/\sqrt{n}}$$

$$= \frac{\bar{X} - E(X_1)}{s_X/\sqrt{n}}$$

$$= t(\mathbf{X}).$$

It should be noted that invariant is defined with respect to the transformation group $\mathcal{G}$, which can be used to characterize invariant transformations.

Definition 3.10 Maximal invariant *A function is maximal invariant if identical mappings imply identical images in the transformation group $\mathcal{G}$.*

The concept of *maximal invariant* can also be formulated as in [81]: A function M is a *maximal invariant* if it is invariant with respect to $\mathcal{G}$, and $M(x_1) = M(x_2)$ implies $x_2 = g(x_1)$ for some $g \in \mathcal{G}$. The concept of maximal invariant can be used to characterize invariant diagnostic predictors as in the following theorem.

Theorem 3.12 *Assume that $M(\mathbf{X})$ is a maximal invariant with respect to a group of transformation $\mathcal{G}$. The necessary and sufficient condition for a diagnostic predictor ϕ to be invariant under $\mathcal{G}$ is that ϕ depends on observations $\mathbf{X}$ only through $M(\mathbf{X})$.*

Proof *A proof of this theorem can be found in [81], where the test statistic serves as the role of a diagnostic predictor.*

Note: In the theorem above, if $x^2 = h(x^3)$, then $h(t) = t^{\frac{2}{3}}$, and $h(t) = \phi(M^{-1}(t))$ is the explicit form of the transforming function h. The following example illustrates the concept of *maximal invariant* for location transformations.

Example 3.14 *Let $\mathbf{X} = (x_1, ..., x_n)^T$, denote the location transformation group*

$$\mathcal{G} = \{g : g(\mathbf{X}) = (x_1 + c, ..., x_n + c)^T, c \in R\}.$$

Let

$$M(\mathbf{X}) = (x_1 - x_n, ..., x_{n-1} - x_n)^T.$$

Then $M(\mathbf{X})$ is maximal invariant under the location transformation $\mathcal{G}$.

Solution:

$$M(g(\mathbf{X})) = \begin{pmatrix} x_1 + c - (x_n + c) \\ \vdots \\ \vdots \\ x_{n-1} + c - (x_n + c) \end{pmatrix} = M(\mathbf{X})$$

if $M(\mathbf{X}) = M(\mathbf{X}')$,

$$
\begin{pmatrix} x_1 - x_n \\ \vdots \\ \vdots \\ x_{n-1} - x_n \end{pmatrix} = \begin{pmatrix} x_1' - x_n' \\ \vdots \\ \vdots \\ x_{n-1}' - x_n' \end{pmatrix},
$$

$x_i - x_n = x_i' - x_n'$ for $i = 1, \ldots, n - 1$. We have g_c: $x_i \longrightarrow x_i'$, $c = x_n' - x_n$, so that $\mathbf{X} = g_c(\mathbf{X}')$, $M(\mathbf{X})$ is a maximal invariant.

Since data transformation occurs frequently in statistical inference (namely, hypothesis testing and estimation), for the remainder of this section, we shall focus on examining the UMEIDP, uniformly most efficient invariant diagnostic predictor. Efficiency is defined in the sense that the sensitivity is maximized for a given specificity.

3.5.2 Invariance and efficiency

When performing data transformations before prediction analysis, the first step should be the verification of invariance, namely the prediction problem is invariant for the corresponding data transformations. We start with the following definition for invariant property in prediction analysis.

Notice that this section focuses on restrained optimization. As for prediction analysis, the target function for optimization is the sensitivity of the diagnostic predictor, while the restraint is the set of invariant predictors. Once we transform the data, the underlying model of the transformed data may change accordingly, alongside the parameter space that defines the domain of the parameter in the assumed model. Toward this end, we need to set the scene for invariant prediction problems.

Definition 3.11 Invariant prediction problem: *For a set of data with an assumed underlying model governed by a parameter θ. The prediction problem for the healthy population $\theta \in \Omega_0$ versus the diseased population $\theta \in \Omega_1$ is invariant under transformation g if the correspondingly transformed parameter is within the original space for healthy subjects and the original space for the diseased patients, respectively.*

Briefly speaking, the invariant prediction problem is for the invariant of the healthy range and diseased range of the diagnostic measure after data transformation. Mathematically, as clearly described in [81] under the setting of hypothesis testing: Let $\bar{g}$ be the corresponding transformation of parameters. The testing problem is invariant if $\bar{g}$ preserves both Ω_0 and Ω_1, $\bar{g}(\Omega_0) = \Omega_0$, $\bar{g}(\Omega_1) = \Omega_1$. In the following discussions involving invariant diagnostic predictors in this book, we confine the discussion to invariant prediction problems.

Definition 3.12 Invariant predictor: *For invariant prediction problems, a* *predictor $\phi(\mathbf{X})$ is invariant with respect to a transformation g if $\phi(g(\mathbf{X})) = \phi(\mathbf{X})$. Namely the prediction outcome remains unchanged after the data transformation.*

With the setting above, the prediction question now becomes to find the restricted optimal solution for invariant predictors (which is the same as the uniformly most efficienty and invariant predictor).

We start with the location transformation for the discussion. Let $\mathbf{X} = (X_1, ...X_n)^T$ be an observation from a population with model $f(x_1, ...x_n)$. Assume that we are interested in predicting the Healthy population defined as

$$f_\theta(x_1, ...x_n) = f_0(x_1 - \theta, ..., x_n - \theta)$$

versus the Diseased population defined as

$$f_\theta(x_1, ...x_n) = f_1(x_1 - \theta, ..., x_n - \theta),$$

for $\theta \in R$. The goal of finding a UMEP invariant predictor is tantamount to finding the one that has uniformly most sensitive, within the set of invariant predictors with respect to a transformation group $\mathcal{G}$.

Note: the problem of predicting diseased or healthy population here is invariant under the group $\mathcal{G}$ of location transformation

$$g(\mathbf{X}) = (X_1 + c, ..., X_n + c),$$

$c \in R$. The corresponding transformation of the model parameter

$$\bar{g}(\theta) = \theta + c, \quad \theta \in \Theta_0$$
$$\Rightarrow f_\theta(\mathbf{X}) = f_0(x_1 + \theta, ..., x_n + \theta), \quad \theta \in R.$$

$$\bar{g}(\theta) \in \bar{g}(\Theta_0)$$
$$\Rightarrow f_{\theta'}(\mathbf{X}) = f_0(x_1 + \theta + c, ..., x_n + \theta + c)$$
$$= f_0(x_1 + c^*, ..., x_n + c^*), \quad c^* \in R.$$

Thus, $\bar{g}(\Theta_0) = \Theta_0 = \{f_0\}$, similarly $\bar{g}(\Theta_1) = \Theta_1 = \{f_1\}$.

Now, consider a maximal invariant under $\mathcal{G}$:

$$S(\mathbf{X}) = \begin{pmatrix} X_1 - X_n \\ \vdots \\ \vdots \\ X_{n-1} - X_n \end{pmatrix} \equiv \begin{pmatrix} t_1 \\ \vdots \\ \vdots \\ t_{n-1} \end{pmatrix},$$

the distribution of $S(\mathbf{X})$ is $f_S(t_1, ..., t_{n-1})$.

Let $t_i = x_i - x_n$, $i = 1, ..., n - 1$, the notation can be simplified as

$$t_n = x_n, \quad \mathbf{T} = (t_1, ..., t_n)^T = (S, T)^T.$$

We have

$$f_{\mathbf{T}}(t_1, ..., t_n) = f_{\mathbf{X}}(x_1(t), ..., x_n(t))|\mathbf{J}|$$
$$= f_{\mathbf{X}}(t_1 + t_n, ..., t_{n-1} + t_n, t_n) \quad |\mathbf{J}| = 1.$$

so, $f_S(t_1, ..., t_{n-1}) = \int_R f_{\mathbf{T}}(t_1, ..., t_n) dt_n$

$$= \int_R f_{\mathbf{X}}(t_1 + t_n, ..., t_{n-1} + t_n, t_n) dt_n.$$

Now for the sample $\mathbf{X}$, we have

$$f_S(t_1, ..., t_{n-1}) = \int_R f_{\mathbf{X}}(x_1 + (t_n - x_n), ..., x_{n-1} + (t_n - x_n), t_n) dt_n,$$

let $t_n - x_n = u \Rightarrow t_n = x_n + u$, we have

$$f_S(t_1, ..., t_{n-1}) = \int_R f_{\mathbf{X}}(x_1 + u, ..., x_{n-1} + u, x_n + u) du.$$

Given a set of data $\mathbf{X}$, the invariant prediction problem on diseased or healthy population under the transformation $g(X_i) = g(X_i) + c$ has the maximal invariant S. For a given S, the associated predicting problem based on S becomes

$$Healthy : f_S(t_1, ..., t_{n-1}) = f_0(t_1, ..., t_{n-1})$$
$$Diseased : f_S(t_1, ..., t_{n-1}) = f_1(t_1, ..., t_{n-1}).$$

Note that since the prediction problem is free of θ in this scenario, it becomes a simple prediction problem discussed in the previous sections. By the Neyman-Pearson lemma, the most sensitive predictor (based on the maximal invariant predictors) is

$$\frac{f_1(t_1, ..., t_{n-1})}{f_0(t_1, ..., t_{n-1})} > c$$

$$\Updownarrow$$

$$\lambda = \frac{\int_R f_1(x_1 + u, ..., x_{n-1} + u, x_n + u) du}{\int_R f_0(x_1 + u, ..., x_{n-1} + u, x_n + u) du} > c, \qquad (3.12)$$

so that $P(\lambda > c | H_0) = \alpha$. The predictor

$$\phi(\mathbf{X}) = \begin{cases} 1 & \text{if } \lambda > c \\ 0 & \text{otherwise} \end{cases}$$

is a UMEP invariant predictor on underlying model for location transformation.

Following the preceding example, we now specifically consider diagnostic predictors on population variation for location transformations.

Example 3.15 *Consider the classification of healthy population characterized as $\sigma = \sigma_0$ versus the diseased population characterized as $\sigma = \sigma_1$, $\sigma_1 > \sigma_0$. Assume that the data $\mathbf{X} \sim N(\theta, \sigma^2)$ where θ is unknown. We are interested in finding the UMEP invariant predictor for location transformations, which is tantamount to finding the restricted optimal solution when the target for maximization is the sensitivity function and the restriction is the location invariant diagnostic predictor with specificity $1 - \alpha$.*

Solution: We may use the discussion on (3.12) as follows. First, consider the model densities under the healthy and diseased populations, respectively,

$$f_0(x_1, ..., x_n) = (\frac{1}{\sqrt{2\pi}\sigma_0})^n \exp(-\frac{\sum_{i=1}^n (x_i - \theta)^2}{2\sigma_0^2})$$

and

$$f_1(x_1, ..., x_n) = (\frac{1}{\sqrt{2\pi}\sigma_0})^n \exp(-\frac{\sum_{i=1}^n (x_i - \theta)^2}{2\sigma_1^2}).$$

Now, the UMEP location invariant diagnostic predictor reads,

$$\lambda(\mathbf{X}) > c \iff (\frac{\sigma_0}{\sigma_1})^n \frac{\int_R \exp(-\frac{-\sum_{i=1}^n (x_i+u-\theta)^2}{2\sigma_1^2})du}{\int_R \exp(-\frac{-\sum_{i=1}^n (x_i+u-\theta)^2}{2\sigma_1^2})du} > c$$

$$\iff \frac{\int_R \exp(-\frac{\sum x_i^2 + 2(u-\theta)\sum x_i + n(u-\theta)^2}{2\sigma_1^2})du}{\int_R \exp(-\frac{\sum x_i^2 + 2(u-\theta)\sum x_i + n(u-\theta)^2}{2\sigma_0^2})du} > c^*,$$

where $c^* = (\frac{\sigma_1}{\sigma_0})^n c)$. Thus,

$$\lambda(\mathbf{X}) > c$$

$$\iff \frac{\exp(-\frac{\sum x_i^2 - n\bar{x}^2}{2\sigma_1^2}) \int_R \exp(-\frac{n((u-\theta)^2 + 2\bar{x}(u-\theta) + \bar{x}^2)}{2\sigma_1^2})du}{\exp(-\frac{\sum x_i^2 - n\bar{x}^2}{2\sigma_0^2}) \int_R \exp(-\frac{n((u-\theta)^2 + 2\bar{x}(u-\theta) + \bar{x}^2)}{2\sigma_0^2})du} > c^*$$

$$\iff \exp(-\frac{1}{2}(\frac{1}{\sigma_1^2} - \frac{1}{\sigma_0^2})(\sum x_i^2 - n\bar{x}^2)) \frac{\int_R \exp(-\frac{n(u-\theta+\bar{x})^2}{2\sigma_1^2})du}{\int_R \exp(-\frac{n(u-\theta+\bar{x})^2}{2\sigma_0^2})du} > c^*$$

$$\iff -\frac{1}{2}(\frac{1}{\sigma_1^2} - \frac{1}{\sigma_0^2})(\sum x_i^2 - n\bar{x}^2) > c^{**}$$

$$\iff \sum_{i=1}^n (x_i - \bar{x})^2 > c^{***}.$$

Notice that

$$\sigma_0 < \sigma_1 \iff \frac{1}{\sigma_0^2} > \frac{1}{\sigma_1^2} \iff -\frac{1}{2}(\frac{1}{\sigma_1^2} - \frac{1}{\sigma_0^2}) > 0.$$

So the UMEIDP, uniformly most efficient invariant diagnostic predictor, for the healthy population vs the diseased population is the one with diagnostic region

$$\mathbf{A} = \{\mathbf{X} : \sum (X_i - \bar{X})^2 > c\},$$

for some c under the group transformation $\{g(x) = x + c, \ c \in R\}$. We shall discuss how to determine the constant c as follows.

For any diagnostic predictor with specificity $1 - \alpha$, we have

$$P_{\sigma_0}(\sum_{i=1}^{n}(X_i - \bar{X})^2 > c)$$
$$= P_{\sigma_0}((n - 1)S^2/\sigma_0^2 > c^*)$$
$$= P(\chi_{n-1}^2 > c^*)$$

So the constant corresponding to the restricted optimal solution is the diagnostic region with $c^* = \chi_{n-1}^2(\alpha)$, where

$$P(\chi_{n-1}^2 > \chi_{n-1}^2(\alpha)) = \alpha,$$

and $c = \sigma_0^2 \chi_{n-1}^2(\alpha)$.

This example shows that the usual χ^2 model is actually the model to determine the diagnostic threshold of a UMEIDP, uniformly most efficient invariant diagnostic predictor, with location transformations on the data. Further theory and techniques on similar topics under data transformations can be found in one of the outstanding resources [81].

SUMMARY This chapter discusses the trade off between sensitivity and specificity. Sensitivity is the probability that a diagnostic criterion correctly identifies the case. It is a measurement confined to diseased population or equivalently a conditional probability given that the patient is indeed sick. On the other hand, specificity is the probability that a diagnostic criterion correctly detects healthy patients. It is a measurement restricted to a healthy population.

For a numerical or digital diagnostic standard, when the cutoff value (threshold) for the measurement (such as body temperature) is low, the rate of misclassification is high. For instance, if a patient is classified as having a fever when the body temperature is more than 96 degrees Fahrenheit, more patients will be classified as having a fever. This of course includes misclassified patients. However, when the criterion is set too high, such as 110 degrees Fahrenheit, more patients will be classified as being healthy, resulting another escalation on misclassification rate. Thus, finding the optimal threshold is critical in the prediction and diagnosis process. Increasing sensitivity is usually at the cost of lowering the specificity, and vice versa. Under this scenario, we elucidate the concept of *uniformly most efficient diagnostic predictor* (UMEP), together with theory and procedures searching for the UMEP.

In the case where UMEP does not exist, we expanded the concept of UMEP into UMEDP (unifomly most efficient and decent diagnostic predictor). The latter engraves UMEP with a new concept, *decent diagnostic predictors*, a legitimate requirement that the rate of correct classification should not be lower than the rate of misclassification. Theory, practical procedures and examples are discussed following the definition of UMEDP.

Data transformation is very common in data science, especially in the process of measurement unification for pooled datasets from multiple resources. Consistent interpretations of insightful information related to sensitivity and specificity necessitates a discussion on the invariant property of UMEP predictor for transformed data. We conclude this chapter with a discussion on theory and procedures regarding invariant UMEP predictors for linear functions in data transformation.

4

Bias and Variation Trade-off

This chapter deals with theoretical and fundamental issues of bias versus variation in data science. There are different views on the content and definition of data science. Some claimed that data science is discernible from statistics due to its unique handling of big data and reliance on modern computer techniques. References in this regard can be found in Bell et. al. (2009) [5] and Dhar (2013) [43], among others. On the other hand, current literature also includes claims that statistics itself is data science (see, for example, Brieman (1998) [11] and Wu (1986) [125], to list just a few). Although various definitions have their own rationales under different scenarios, in our view, the essential process of data science is to make inference, to predict (or forecast) the unknown using the known (observable data). Thus, without confining ourselves into either direction, from the viewpoint of data analytic technologies, algorithms, and methodological development, we go with the belief that data science includes the model-based camp (mainly statistics) and the data-driven camp (mainly computing techniques, machine learning, and deep learning algorithms), as elucidated in Chapter One. In the process of predicting the unknown, one frequently asked issue focuses on the dilemma regarding the bias and variation of the predicted outcome: gaining lower bias at the cost of high variation or trading prediction bias for lower prediction variation.

4.1 Reducible and Irreducible Errors in Prediction

Discussions on data science often include two fundamental and integrated parts, existing statistical methodologies (such as estimation, hypothesis testing, and prediction), as well as data-driven methods (such as neural networks and deep learning). As introduced in Chapter One, in both camps of data science, optimization is ubiquitous in the development of data analytic procedures. Optimization strategies cover from the minimization of the residual network errors in deep learning to the selection of the uniformly most powerful test in hypothesis testing for model-based inference. However, the optimal solution does not always exist for practical problems.

When the optimal solution does not exist, we need to reformulate the analytical problem and seek local optimization. This necessitates a discussion on optimization issues within a confined domain. At this point, it is convenient to unify various analytical procedures into a general and fundamental framework, which is the concept of restricted optimization as defined below.

Definition 4.1 Restricted Optimization: *Let (Y, X) be a data set with the underlying relationship*

$$Y = h(X) + \epsilon.$$

Assume that the evaluation criterion is $\mathcal{F}$, the set of all permissible function for $h(.)$ is Γ, and Δ is a subset of Γ. The process of finding $\hat{h}$, such that

$$\mathcal{F}(\hat{h}) = arg \inf_{h \in \Delta} \mathcal{F}(h); \quad or \quad \mathcal{F}(\hat{h}) = arg \sup_{h \in \Delta} \mathcal{F}(h),$$

is a restricted optimization process, in which $\mathcal{F}$ is the target function and the set Δ is the restriction.

Under this definition, methods in hypothesis testing, estimation, and prediction can be reformulated into the framework of restricted optimization. For instance, finding UMPU (UMP unbiased) test is a process in which we maximize the power of the the test with restriction to unbiased tests. Also, finding the best linear prediction can be viewed as a restricted optimization in which we minimize the expected prediction error with a restriction to linear functions of features involved in prediction.

It should be noted that the framework of restricted optimization includes machine learning in which it uses the duality theorem in linear programming to identify a support vector machine. It also includes deep learning when convolution over a specified set is applied (to list just a few).

The key component in the definition of restricted optimization is the target function $\mathcal{F}$, which determines the trained model in the data validation stage. It also affects the selection of the evaluation criterion at the stage where we apply the testing data for prediction errors.

It is related at this point to clarify the concept of expected prediction error for a given set of data. Recall the general practice in data science (where a set of big data is available), we usually split the data into 75:25. Namely 75% of the original data is treated as the training data, and the remaining 25% is the testing data. During the training process, we want to minimize the expected prediction error. The process of minimization in this setting is an optimization process. After training, the trained model is then evaluated by the testing data set to examine the possibility of over-fitting (incorrectly including patterns of random effects contaminated in the training data). In this setting, the target function is the expected prediction error (EPE), defined in Chapter One.

Let $\hat{Y}$ be the prediction of Y. One of the commonly applied criteria for prediction is the expected prediction error (EPE),

$$
\begin{aligned}
E[(Y - \hat{Y})^2] &= E\{[f(X) + \epsilon - \hat{f}(X)]^2\} \\
&= E[f(X) - \hat{f}(X)]^2 + var(\epsilon). \qquad (4.1) \\
&= reducible\ error + irreducible\ error
\end{aligned}
$$

FIGURE 4.1
Statistics versus Data Science

Obviously, the above expression suggests that EPE can be decomposed into two portions. One accounts for the error between the predicted underlying model and the true model, which is reducible when we have large enough training data in conjunction with legitimate features for prediction in the data. Another portion accounts for the error of the randomness of the data, $var(\epsilon)$, which is due to the intrinsic fluctuation of the data that we cannot influence in prediction, and is irreducible.

To further examine the evaluation of the expected prediction error, consider a scenario in which we only have finite distinguishable features and responses (y_j, X_j) for $j = 1, ..., p$.

Now, when we have a set of testing data (y_i, X_i) with $i = 1, ..., m$, and $\hat{f}(X)$, a model learned from the training data (y_i, X_i) for $i = m+1, ..., n$, the sample expected prediction error (of the testing data) reads

$$\frac{1}{m}\sum_{k=1}^{m}[y_k - \hat{f}(X_k)]^2 = \sum_{j=1}^{p}[y_j - \hat{f}(X_j)]^2 \frac{freq((y_j, X_j))}{m},$$

As the amount of test data goes to infinite, $m \to \infty$,

$$\lim_{m\to\infty} \frac{freq(y_j, X_j)}{m} = P[(Y, X) = (y_j, X_j)].$$

Thus,

$$\frac{1}{m}\sum_{k=1}^{m}[y_k - \hat{f}(X_k)]^2 \to E[(Y - \hat{Y})^2],$$

the sample prediction error approaches the expected prediction error when the sample size of the testing data is large enough. This shows that although we need to reserve a good portion of data to train the model, we also need to keep a good size of data for the testing data. If the size of the testing data is too small, over-estimating or under-estimating the expected prediction error may lead to a misleading conclusion on the performance of the trained model.

Assume that the testing set contains large enough observations. When the solution to the global optimization is available, the reducible error in (4.1) is minimized and the expected prediction error cannot be further reduced. However, when the global optimization solution does not exist, the analytic process stops. In what follows, we shall discuss details of the fundamental concepts and terminologies used in statistical prediction and machine learning, when the global optimization needs to be confined with restrictions.

We will start with examples of restricted optimization when the underlying model is assumed (model-based inference), which is followed by a discussion on the impact of nuisance parameters. After that, we will discuss restricted optimizations for fundamental estimation issues in data transformation, including the invariant property and location-scale transformations. For transformed data, a representative topic on restricted optimization is the minimization of the variance confined to unbiased estimators in model-based inference.

Certainly, it is impossible to exhaust all restricted optimization methods in one chapter. To cover key issues in restricted optimization, we merely focus on underpinning ideas and representative principles. Discussions in this chapter may help clarify the premises of algorithms in data science (to avoid the abuse of data analytic procedures), enhancing understanding of prediction procedures, and facilitating interpretations of analytical outcomes.

Materials in the rest of this chapter also underpin rationales and theory behind common statistical decisions. For example, the Student-t test can actually be viewed as an optimal test in terms of maximizing the power of the test with restriction to unbiased tests, when nuisance parameters are involved. Most of the results presented in this chapter are synthesized from theorems in classical textbooks such as Lehmann and Romano [81], Lehmann and Casella [80], Casella and Berger [16], Shao [110], and Hastie et al (2009) [56], among others. The revisit in this chapter sheds new light on classical results for restricted optimization in data science.

4.2   Minimum variance unbiased estimators

The preceding section focuses on optimization on power for hypothesis testing and risk function for estimation with restriction on data transformations. In this section, we shall discuss another strategy of restricted optimization, in which the target function of optimization is to minimize the variance of the estimator with the restriction on being an unbiased estimator. We start with the definition.

Definition 4.2 Uniformly minimum variance unbiased estimator (UMVUE): *For a set of data with an unknown parameter that characterizes the underlying model. Consider the estimator (function of the data) for the unknown parameter in a permissible set. If an estimator (with finite second moment) is unbiased (the long-term average of the estimator hits the parameter), and it reaches the lowest possible variance among all the unbiased estimators uniformly for all permissible values of the parameter, such an estimator is called uniformly minimum variance unbiased estimator, or UMVUE.*

As documented in many advanced statistics textbooks (such as [80], [16], among others), the concept of UMVUE can be described mathematically as follows. Consider the estimation of a function of parameter $g(\theta)$ based on a set of data $\mathbf{X} = (X_1, \ldots\ldots, X_n)$ from a family of distribution P_θ, $\theta \in \Omega$, where Ω is the parameter space. An unbiased estimator $\delta(\mathbf{X})$ of $g(\theta)$ is UMVUE if for every $\theta \in \Omega$,

$$Var(\delta(\mathbf{X})) \leq Var(\delta'(\mathbf{X})),$$

where Var denotes the variation of a random variable, and $\delta'(\mathbf{X})$ is another unbiased estimator of $g(\theta)$.

Based on the above definition, any estimator is unbiased in its own expectation. The following theorem serves as an illustrating example of the restricted optimization process in the construction of a UMVUE. The feature of this restricted optimization process is that the UMVUE of its own expectation can be identified by a condition with the unbiased estimator of zero, without minimizing the target function directly.

Theorem 4.1 *Denote the underlying model of a set of data $\mathbf{X}$ by P_θ, where $\theta \in \Omega$ is the parameter characterizing the underlying model. Assume that δ is an estimator that has finite second moment (so that it is meaningful to discuss its variance),*

$$E_\theta(\delta^2) < \infty.$$

A necessary and sufficient condition for δ to be a UMVUE of its expectation is that

$$E_\theta(\delta U) = 0, \tag{4.2}$$

for all $U \in \mathcal{H}$ and any $\theta \in \Omega$, where $\mathcal{H}$ is the set of all unbiased estimators of zero which finite second moment.

Proof *We start with the necessity part of the theorem. Assume that δ is the UMVUE of its expectation and denote*

$$E_\theta(\delta) = g(\theta).$$

For any $U \in \mathcal{H}$ and $\theta \in \Omega$, and for an arbitrary real value λ, denote

$$\delta' = \delta + \lambda U,$$

obviously, δ' is another unbiased estimator of $g(\theta)$. Consider

$$Var_\theta(\delta + \lambda U) \geq Var_\theta(\delta)$$

for all λ. Expanding the left-hand side of the above equation, we have

$$\lambda^2 Var_\theta(U) + 2\lambda Cov_\theta(\delta, U) \geq 0$$

for all λ. This is a quadratic equation of a real value λ with two roots $\lambda_1 = 0$ and

$$\lambda_2 = -2Cov_\theta(\delta, U)/Var_\theta(U).$$

It will therefore take negative values unless

$$Cov_\theta(\delta, U) = 0,$$

which implies (4.2).

As for the sufficiency part of the theorem, suppose (4.2) is valid for all $U \in \mathcal{H}$. To show that δ is UMVUE of its expectation, let δ' be another unbiased estimator of $E_\theta(\delta)$. If $Var_\theta(\delta') = \infty$, there is no need to prove since its variance is larger than that of δ. So, we can assume $Var_\theta(\delta') < \infty$. In this case,

$$\delta' - \delta \in \mathcal{H}$$

because they are both unbiased estimators of $E(\delta)$. Furthermore,

$$E_\theta[\delta(\delta' - \delta)] = 0,$$

hence

$$E_\theta(\delta'(\delta)) = E_\theta(\delta^2).$$

Now, using the fact that δ' and δ have the same expectation, we have

$$Var_\theta(\delta) = Cov(\delta, \delta') \leq \sqrt{Var_\theta(\delta)Var_\theta(\delta')},$$

thus,

$$Var_\theta(\delta) \leq Var_\theta(\delta').$$

The beauty of the above theorem links the property of a statistic with the restricted optimal solution in terms of minimizing the variance (or the risk under the square loss function). For instance, if the distribution family is complete, which means that for any statistic U, $E_\theta(U) = 0$ implies that $U = 0$ almost surely. The above theorem points out an important fact in mathematical statistics that any unbiased and complete estimator is the UMVUE of its own expectation. We shall illustrate this point with the following example.

Example 4.1 *Let X be a random variable with $E(X^2) < \infty$. For a set of normal data $X_1, ..., X_n$ that have underlying model $N(0, \sigma^2)$, find the UMVUE of the population standard deviation σ.*

Solution: First, notice that $S = \sqrt{\frac{1}{n-1}\sum_{i=1}^{n}(X_i - \bar{X})^2}$, the sample standard deviation, is not unbiased for $\sigma = \sqrt{Var(X)}$. To see this point, considering the normality assumption, we have

$$Y = \frac{n-1}{\sigma^2}S^2 \sim \chi_{n-1}^2,$$

and

$$E(S) = E(\frac{\sigma}{\sqrt{n-1}}\sqrt{Y})$$
$$= \frac{\sigma}{\sqrt{n-1}}\int \sqrt{y}f_Y(y)dy$$
$$= \sqrt{\frac{2}{n-1}}\frac{\Gamma(\frac{n}{2})}{\Gamma(\frac{n-1}{2})}\sigma.$$

which is not σ. In general, we have

$$E(S^2) - (E(S))^2 = Var(X) \geq 0.$$

Now,

$$E(S) \leq \sqrt{E(S^2)} = \sigma,$$

since

$$E(S^2) = E(\frac{1}{n-1}\sum_{i=1}^{n}(X_i - \bar{X})^2) = \sigma^2.$$

For the model of the data in this example, $S^2 = \sum_{i=1}^{n} X_i^2$ is a complete statistic for σ, and $\frac{S^2}{\sigma^2} \sim \chi_n^2$, thus,

$$E(\frac{S^r}{\sigma^r}) = \frac{2^{\frac{r}{2}}\Gamma(\frac{n+r}{2})}{\Gamma(\frac{n}{2})}$$

for any positive integer r. Therefore, by Theorem 4.1, the UMVUE of σ reads

$$\hat{\sigma} = \frac{\Gamma(\frac{n}{2})}{\sqrt{2}\Gamma(\frac{n+1}{2})}S.$$

This example shows that the restricted optimization for the estimation of the standard deviation that governs the model behind a set of data is not the sample standard deviation. Although the sample variance is an unbiased estimator of the population variance, the best estimator of the sample standard deviation is actually the sample standard deviation multiplied by a non-unit constant.

We shall now discuss another example which is related to the application of Theorem 4.1 when the underlying model of the data is a discrete model. Another related issue in this regard is the existence of the UMVUE for a function of the unknown parameter. The following example shows how to use Theorem 4.1 to determine whether the UMVUE of a parameter exists, and how to obtain UMVUE for discrete data.

Example 4.2 *Let X take on the values -1, 0, 1, 2, 3 with probabilities $P(X = -1) = 2pq$ and $P(X = k) = p^k q^{3-k}$ for k =0, 1, 2, 3, where $0 < p < 1$, $q = 1 - p$.*

(a) Verify that $P(X = k)$ given above is a probability model.

(b) Determine whether the UMVUE of p exists, given one observation X. How about the UMVUE of pq?

Solution: As for part (a), notice that

$$\sum_{k=-1}^{3} P(X = k) = 2pq + q^3 + p^2q + pq^2 + p^3$$

$$= (1-p)^3 + p^2(1-p) + p(1-p)^2 + p^3 + 2p(1-p)$$
$$= (1-p)^2(1-p+p) + p^2(1-p) + p^3 - 2p(1-p)$$
$$= (1-p)^2 + p^2 - 2p(1-p)$$
$$= p^2 + 2pq + q^2$$
$$= (p+q)^2$$
$$= 1$$

For part (b), according to Theorem 4.1, we need to find an estimator $\delta(X)$ such that $E[\delta(X)] = 0$ and check the existence of the unbiased estimator $T(X)$ such that $E[\delta(X)T(X)] = 0$ for all $\delta(X)$. If such an unbiased estimator exists, it is UMVUE; otherwise, the UMVUE does not exist. Towards this end, for any unbiased estimator, δ, of the value 0 in the setting of this problem, we have

$$E[\delta(X)]$$

$$= \sum_{k=-1}^{3} \delta(k)P(X = k)$$

$$= p^3[\delta(3) - \delta(2) + \delta(1) + \delta(0)] + p^2[\delta(2) - 2\delta(1) + 3\delta(0) - 2\delta(-1)]$$
$$+ p[\delta(1) - 3\delta(0) + 2\delta(-1)] + \delta(0)$$

$$= 0.$$

This is a polynomial of p, we want the equality to hold for every p. Setting the coefficients to 0, we have

$$\delta(0) = 0, \delta(2) = \delta(1) = -2\delta(-1), \delta(3) = 0,$$

which can be simplified as

$$\delta(-1) = a, \delta(-2) = \delta(1) = -2a,$$

for some constant $a \in R^+$. Similarly, for any unbiased estimator of p, $T(X)$, if $E(T(X)) = p$, we have

$$T(0) = 0, T(3) = 1, \text{ and } T(1) + 2T(-1) = 1.$$

Now, if

$$
\begin{aligned}
&E[T(X)\delta(X)] \\
&= 2ap^3[T(2) - T(1)] + 2ap^2[T(1) - T(2)] + 2ap[T(-1) - T(1)] \\
&= 0,
\end{aligned}
$$

for every δ, $T(X)$ is the UMVUE of its expectation if and only if

$$T(-1) = T(1) = T(2) = b$$

for some constant $b \in R$. Plugging the expression on b into the expression of $E[T(X)]$ yields

$$E[T(X)] = p^3 - 3bp^2 + 3bp = p, \text{ and } b = \frac{1+p}{3}.$$

As a result, if the UMVUE exists, the estimator $T(X)$ has to be a function of the unknown parameter p. This is in contradiction with the definition of a statistic (which is a function of the data only, not involving the unknown parameter). Thus, the UMVUE of p doesn't exist for the model in this question.

However, when we consider the UMVUE of a function of the parameter pq, if

$$
\begin{aligned}
&E[T(X)] \\
&= 2pqT(-1) + q^3T(0) + pq^2T(1) + p^2qT(2) + T(3)p^3 \\
&= pq,
\end{aligned}
$$

we have

$$T(3) = T(0) = 0, \text{ and } b = \frac{1}{3}.$$

In this case, the UMVUE for pq exists.

The above theorem obtains the restricted optimization by means of an unbiased estimator of zero, although the optimization process is hidden in the proof of the theorem. In what follows, we shall discuss a well-known theorem that directly approaches the minimal value of the target function (the variance) to find the UMVUE.

Theorem 4.2 Cramer-Rao optimization: *Assume that the underlying model for the data* $\mathbf{X} = (X_1, ...X_n)$, $f(\mathbf{X}|\theta)$, *satisfies the Leibniz condition for limits and integrations. Denote* $W(\mathbf{X}) = W(X_1, ...X_n)$, *an estimator satisfying*

$$\frac{d}{d\theta}E_\theta W(\mathbf{X}) = \int_\mathcal{X} \frac{\partial}{\partial\theta}[W(\mathbf{X})f(\mathbf{X}|\theta)]d\mathbf{X}$$

and $Var_\theta(W(\mathbf{X})) < \infty$. *Then*

$$Var_\theta(W(\mathbf{X})) \geq \frac{\frac{d}{d\theta}E_\theta(W(\mathbf{X}))}{E_\theta[(\frac{\partial}{\partial\theta}\log f(\mathbf{X}|\theta))^2]}. \tag{4.3}$$

Note: The above inequality (4.3) points out the lowest possible value for variances of unbiased estimator. If an unbiased estimator (restriction) has the variance equal to the right-hand side of (4.3), it is the UMVUE of θ. The validity of the theorem can be shown as follows. More thorough discussions can be found in [16], [80], among others.

Proof *First, applying the derivative on the unbiased restriction with the use of the Leibniz condition, we have*

$$\frac{d}{d\theta}E_\theta W(\mathbf{X})$$
$$= \int_\mathcal{X} W(\mathbf{X})\frac{\partial}{\partial\theta}f(\mathbf{X}|\theta)d\mathbf{X}$$
$$= E_\theta[W(\mathbf{X})\frac{\frac{\partial}{\partial\theta}f(\mathbf{X}|\theta)}{f(\mathbf{X}|\theta)}] \quad \textit{by multiplying } \frac{f(\mathbf{X}|\theta)}{f(\mathbf{X}|\theta)} \textit{ in the integrand}$$
$$= E_\theta[W(\mathbf{X})\frac{\partial}{\partial\theta}\log f(\mathbf{X}|\theta)]$$

Now, consider the basic property of a density function. We have,

$$E_\theta[\frac{\partial}{\partial\theta}\log f(\mathbf{X}|\theta)] = \frac{d}{d\theta}E[1] = 0,$$

thus

$$Cov_\theta[W(\mathbf{X}), \frac{\partial}{\partial\theta}\log f(\mathbf{X}|\theta)]$$
$$= E_\theta[W(\mathbf{X})\frac{\partial}{\partial\theta}\log f(\mathbf{X}|\theta)]$$
$$= \frac{d}{d\theta}E_\theta W(\mathbf{X}).$$

By the Cauchy-Schwartz inequality:

$$[Cov(X,Y)]^2 \leq Var(X)Var(Y),$$

in conjunction with

$$Var_\theta[\frac{\partial}{\partial\theta}\log f(\mathbf{X}|\theta)] = E_\theta[(\frac{\partial}{\partial\theta}\log f(\mathbf{X}|\theta))^2],$$

we have:

$$Var_\theta(W(\mathbf{X})) \geq \frac{\frac{d}{d\theta}E_\theta(W(\mathbf{X}))}{E_\theta[(\frac{\partial}{\partial\theta}\log f(\mathbf{X}|\theta))^2]}.$$

The following example shows how to use the Cramer-Rao lower bound to find the UMVUE (minimizing variance with restriction to unbiased estimators).

Example 4.3 *Assume that the underlying model of a set of data $X_1, ..., X_n$ is the exponential model with unknown parameter λ, exp(λ), find the UMVUE of λ.*

Solution: For an exponential model, we know

$$E(X_1) = \lambda, \ Var(X_1) = \lambda^2,$$

and

$$E(\bar{X}) = \lambda,$$

where $\bar{X}$ is the sample mean. Also,

$$\log f(X_1) = -\log\lambda - \frac{X_1}{\lambda},$$

and

$$\frac{\partial}{\partial\lambda}\log f(X_1) = -\frac{1}{\lambda} + \frac{X_1}{\lambda^2}.$$

Thus,

$$E[(\frac{\partial}{\partial\lambda}\log f(X_1))^2] = E[(\frac{X_1}{\lambda^2} - \frac{1}{\lambda})^2]$$
$$= \frac{1}{\lambda^2} - \frac{2E(X_1)}{\lambda^3} + \frac{E(X_1^2)}{\lambda^4}$$
$$= \frac{1}{\lambda^2}.$$

In this case, the C-R lower bound is $\frac{\lambda^2}{n}$, which implies that the UMVUE of λ is $Var(\bar{X})$.

Following the Cramer-Rao lower bound on the variance of unbiased estimators, another technique in the process of restricted optimization is the approach of approaching the optimal solution by conditioning on an sufficient statistic.

Theorem 4.3 Rao-Blackwell theorem: *For a set of data* $\mathbf{X} = (X_1, \ldots, X_n)$, *assume that* $W(\mathbf{X})$ *is any unbiased estimator of a function of parameters* $\tau(\theta)$, *and let* $T(\mathbf{X})$ *be a sufficient statistic for the unknown* θ. *Define* $\phi(T) = E(W|T)$, *then* $\phi(T)$ *is also unbiased with variance smaller than the variance of* W *for every* θ.

Proof *Obviously*

$$E(\phi(T)) = E[E(W|T)] = E(W),$$

thus, $\phi(T)$ *is unbiased for* $E(W)$, *also:*

$$\begin{aligned} Var_\theta(W) &= Var_\theta[E(W|T)] + E_\theta[Var_\theta(W|T)] \\ &= Var_\theta[\phi(T)] + E_\theta[Var_\theta(W|T)] \\ &\geq Var_\theta[\phi(T)] \end{aligned}$$

Because the sufficiency of T, *the conditional distribution of* $\mathbf{X}$ *given* T, *does not depend on* θ, $\phi(T)$ *is indeed a statistic.*

The Rao-Blackwell theorem indicates that if T is a complete statistic for θ, then any estimator based on T is the best UMVUE of its expected value.

Example 4.4 Rao-Blackwellization: *Suppose that* X_i, $i = 1, \ldots, n$ *are Bernoulli(p), for* $n \geq 4$. *Show that the product* $X_1 X_2 X_3 X_4$ *is an unbiased estimator of* p^4, *and use this fact to find the UMVUE of* p^4

Solution: It is natural to think about $\bar{X}$ since it is a sufficient and complete statistic for p. However, $\bar{X}^4$ is not an unbiased estimator for p^4. To see this point, notice that,

$$\begin{aligned} E[(\sum_{i=1}^n X_i)^4] &= E(\sum_i \sum_j \sum_k \sum_l X_i X_j X_k X_l) \\ &= \sum_i E(X_i^4) + \sum_{i \neq j} E[(X_i X_j)^2] \\ &\quad + \sum_{i \neq j \neq k, l = i, j, \text{or} k} E[(X_i X_j X_k) X_l] + \sum_{i \neq j \neq k \neq l} E(X_i X_j X_k X_l) \\ &= np + \binom{n}{2} p^2 + \binom{n}{3} p^3 + \binom{n}{4} p^4 \end{aligned}$$

Thus, $\bar{X}^4$ is not an unbiased estimator for p^4. In fact, the UMVUE can be constructed by using the Rao-Blackwell approach as stated in the previous theorem.

By independence,

$$E(X_1 X_2 X_3 X_4) = \prod_{i=1}^{n} X_i = p^4,$$

so the estimator $S = X_1 X_2 X_3 X_4$, is unbiased for p^4.

Now, since the sample summation $T = \sum_{i=1}^{n} X_i$ follows a binomial model, which is a complete statistic,

$$\phi(T) = E(X_1 X_2 X_3 X_4 | T)$$

is the UMVUE of its expected value.

Notice that $X_1 X_2 X_3 X_4$ only takes values 0 or 1. We can find the conditional expectation as follows.

For $4 \leq T \leq n$, we have

$$\begin{aligned}
\phi(T) &= E(X_1 X_2 X_3 X_4 | T = t) \\
&= \frac{P(X_1 = X_2 = X_3 = X_4 = 1, \sum_{i=5}^{n} X_i = t - 4)}{P(T = t)} \\
&= \frac{p^4 \binom{n-4}{t-4} p^{t-4}(1-p)^{n-t}}{\binom{n}{t} p^t (1-p)^{n-t}} \\
&= \frac{\binom{n-4}{t-4}}{\binom{n}{t}}.
\end{aligned}$$

For $T < 4$, $\phi(T) = 0$.

Thus, denote the sample summation as T, the UMVUE for p^4 is

$$\frac{\binom{n-4}{T-4}}{\binom{n}{t}}$$

when $T \geq 4$; and it is zero when $T < 4$.

The above three theorems discuss three different approaches in restricted optimization when we are using the square loss function (variance) in estimation with restriction to unbiased estimators. It should be noted that the restricted optimal solution is not always the most accurate and robust estimator, as shown in the following example.

Example 4.5 Small sample estimation: *Given a random sample $X_1, \ldots\ldots X_n$ from the Poisson family with unknown parameter θ, find the UMVUE for $g(\theta) = e^{-10\theta}$*

Solution: Notice that in this setting, $T = \sum_{i=1}^{n} X_i$ is complete and follows a

Poisson distribution with parameter $n\theta$. For the UMVUE, we want $\delta(T)$ to be unbiased, which means that

$$E[\delta(T)] = \sum_{t=0}^{\infty} \delta(t) \frac{(n\theta)^t}{t!} e^{-n\theta} = e^{-10\theta},$$

thus

$$\sum_{t=0}^{\infty} \delta(t) \frac{(n\theta)^t}{t!}$$

$$= e^{(n-10)\theta}$$

$$= \sum_{t=0}^{\infty} \frac{(n-10)^t \theta^t}{t!},$$

comparing each term gets

$$\delta(T) = (1 - \frac{10}{n})^T.$$

However, when $n = 1$, $\delta(T) = (-9)^{X_1}$, the value of UMVUE oscillates wildly between positive and negative values. It is not even close to $g(\theta)$.

This example shows that although UMVUE is optimal in the sense of minimizing the variance for unbiased estimators, it may perform poorly in estimating/predicting the true value of the unknown parameter in the scenario discussed in this example. When the sample size is not large enough, enhancing the condition of unbiasedness may result in poor performance of the estimation. In the next section, we shall discuss another aspect on restricted optimization in which the minimum variance unbiased restriction is replaced by the restriction of minimum risk estimator.

4.3   Minimum risk estimators for transformed data

Another aspect of restricted optimization in model-based inference focuses on estimation. We shall discuss the idea of minimum risk invariant estimator and minimum risk equivariant estimator for transformed data in this subsection. We start with three basic notations that will be used frequently in the sequel.

1 A data transformation defined on the sample space $g : \Omega \to \Omega$.

2 For each data transformation, the distribution of the transformed data changes from the original P_θ to $P'_{\theta'}$. Correspondingly, there is a change of the model parameter: $\bar{g} : \mathcal{H} \to \mathcal{H}'$, so that

$$\theta' = \bar{g}(\theta).$$

3 For each g and $\bar{g}$ the corresponding change of the value of parameter for estimation: g^*: $h(\theta') = h(\bar{g}(\theta)) = g^*(h(\theta))$.

Note: If $G = \{g\}$ is a group, then $\bar{G} = \{\bar{g}\}$ and $G^* = \{g^*\}$ are groups. Also, G, $\bar{G}$ and G^* are *isomorphic*,

$$P_\theta(g(X) \in A') = P_{\bar{g}(\theta)}(X \in A), \quad \text{or } E_\theta[\Psi(gX)] = E_{\bar{g}\theta}[\Psi(X)].$$

Analogically as maximizing the power function in hypothesis testing for data analysis, in statistical estimation, the target for optimization is to minimize the expectation of the loss function. We shall now review the concept of invariant loss function upon data transformations.

Definition 4.3 (Invariant estimation problem) *Assume that the probability model $\mathcal{P} = \{P_\theta, \theta \in \Omega\}$ is invariant under a transformation g. Denote the associated transformation on the parameter as $\bar{g}$ and the associated transformation on the estimate as g^*. Assume that the loss function is invariant after the transformation, namely, L satisfies*

$$L(\bar{g}\theta, g^*d) = L(\theta, d),$$

for any parameter θ and estimate d. Further, assume that the corresponding function of parameter to be estimated, $h(\theta)$, satisfies

$$h(\theta_1) = h(\theta_2) \Rightarrow h(\bar{g}\theta_1) = h(\bar{g}\theta_2).$$

Under this setting, the problem of estimating $h(\theta)$ with loss function L is invariant under g.

For invariant estimation problem, since an estimator is essentially a function of the data, we may define the concept of equivariant estimator as follows. Further discussion on this topic can be found in [80], or [16].

Definition 4.4 (Equivariant estimator) *In an invariant estimation problem, an estimator $\delta(\mathbf{X})$ is said to be* equivariant *if the estimated value based on the transformed data equals the value of the corresponding transformation of the estimator based on the original data. Mathematically, it can be expressed as*

$$\delta(g\mathbf{X}) = g^*\delta(\mathbf{X}).$$

With the above definitions on the estimation problem for transformed data, we have the following theorem. It states that for an invariant estimation problem, the risk of an equivariant estimator is invariant.

Theorem 4.4 *If δ is an equivariant estimator in an invariant estimation problem under a transformation g, the risk function of δ satisfies*

$$R(\bar{g}\theta, \delta) = R(\theta, \delta).$$

Proof *By definition,* $R(\bar{g}\theta, \delta) = E_{\bar{g}\theta}L[\bar{g}\theta, \delta(\mathbf{X})]$, *the right side is equal to*

$$E_\theta L[\bar{g}\theta, \delta(g\mathbf{X})] = EL[\bar{g}\theta, g^*\delta(\mathbf{X})] = R(\theta, \delta),$$

since the loss function is invariant $L(\bar{g}\theta, g^*d) = L(\theta, d)$.

Note: According to the above theorem, if $\bar{G}$ is transitive over the parameter space Ω, the risk function of any equivariant estimator is a constant, being independent of θ. This theorem lays the legitimacy for the definition of the minimum risk equivariant estimator.

Definition 4.5 (Minimum risk equivariant estimator) *In an invariant estimation problem, if there exists an equivariant estimator that minimizes the constant risk, such estimator is called the* minimum risk equivariant *(MRE) estimator.*

We shall now apply the above theoretical concepts to two frequently used groups in data transformations. One is for location transformation and another is scale transformation. We start with the location transformation first.

Example 4.6 *A family of densities* $f(x|\theta)$ *with parameter* θ *and loss function* $L(\theta, \delta)$ *is a* location invariant model *if* $f(x'|\theta') = f(x|\theta)$.

Notice that $L(\theta, \delta)$ is a *location invariant loss function* if $L(\theta, \delta) = L(\theta', \delta')$, where the transformed data

$$x' = x + a,$$

correspond to the transformed parameter

$$\theta' = \theta + a$$

and the estimated value

$$\delta' = \delta + a$$

for any location shift value a.

For example, an invariant loss function for location transformation reads,

$$L(\xi, d) = \rho(d - \xi). \tag{4.4}$$

An estimating problem is location invariant if both family and loss function are invariant.

Now, assume that the underlying model for a set of data $\mathbf{X} = (X_1, ..., X_n)$ takes the following format,

$$f(\mathbf{X} - \xi) = f(x_1 - \xi, ..., x_n - \xi), \quad -\infty < \xi < +\infty, \tag{4.5}$$

where f is known and ξ is an unknown location parameter. Suppose that for the problem of estimating ξ with loss function $L(\xi, d)$ as in (4.4), we have an estimator $\delta(\mathbf{X})$.

The above setting lays the foundation for the discussion of restricted optimization with the target to minimizing the risk of the estimator and the restriction on equivariant estimators, for location transformations of the original data. First, we shall identify the set of equivariant estimators for location transformations.

Theorem 4.5 *If δ_0 is any equivariant estimator in an invariant estimation problem, a necessary and sufficient condition for any estimator δ to be an equivariant estimator is*

$$\delta(\mathbf{X}) = \delta_0(\mathbf{X}) + u(\mathbf{X}),$$

where u is a function satisfying

$$u(\mathbf{X} + a) = u(\mathbf{X}) \tag{4.6}$$

for all data $\mathbf{X}$ and real value a.

Theorem 4.6 *Assume that the underlying model for a set of data $\mathbf{X}$ takes the format in (4.5), and δ is equivariant for estimating ξ with loss function (4.4). Then, the bias, risk, and variance of δ are all constant, being independent of ξ.*

Notice that a function u satisfies (4.6) if and only if it is a function of the differences $y_i = x_i - x_n$ ($i = 1, ..., n-1$) when $n \geq 2$, and if and only if it is a constant when $n = 1$. Thus, the above theorem can be expressed as follows.

Theorem 4.7 *If δ_0 is any equivariant estimator, a necessary and sufficient condition for an estimator δ to be equivariant is that there exists a function v of $n - 1$ arguments for which*

$$\delta(\mathbf{X}) = \delta_0(\mathbf{X}) - v(\mathbf{y}) \tag{4.7}$$

With the above preparation, the following theorem constructs an optimal solution on minimizing risk with restriction to equivariant estimators in data transformation.

Theorem 4.8 *Assume that the underlying model for a set of data $\mathbf{X}$ is (4.5). Let $Y_i = X_i - X_n$ for $i = 1, ..., n-1$ and denote $\mathbf{y} = (Y_1, ..., Y_{n-1})$. Suppose the loss function is given by (4.4), and there is an equivariant estimator δ_0 of ξ with finite risk. Also assume that for each $\mathbf{y}$, there exists a number $v(\mathbf{y}) = v^*(\mathbf{y})$ which minimizes*

$$E_0\{\rho[\delta_0(\mathbf{X}) - v(\mathbf{y})]|\mathbf{y}.\} \tag{4.8}$$

Then a location equivariant estimator with minimum risk exists, and is given by

$$\delta^*(\mathbf{X}) = \delta_0(\mathbf{X}) - v^*(\mathbf{y}).$$

Bias and Variation Trade-off

Proof *By Theorem 4.7, the MRE estimator is found by determining v so as to minimize*

$$R_\xi(\delta) = E_\xi\{\rho[\delta_0(\mathbf{X}) - v(\mathbf{y}) - \xi]\}.$$

Since the risk is independent of ξ, it is suffices to minimize

$$R_0(\delta) = E_0\{\rho[\delta_0(\mathbf{X}) - v(\mathbf{y})]\}$$

$$= \int E_0\{\rho[\delta_0(\mathbf{X}) - v(\mathbf{y})]|\mathbf{y}\}f(\mathbf{y})dy. \tag{4.9}$$

The integral is minimized by minimizing the integrand, and hence (4.9) for each $\mathbf{y}$. Since δ_0 has finite risk $E_0\{\rho[\delta_0(\mathbf{X})]|\mathbf{y}\} < \infty$, the minimization of (4.9) is meaningful. The result now follows from the assumption of the theorem.

The following examples illustrate the construction of an MRE estimator for location transformations.

Example 4.7 Pitman Estimator: *Under the assumption of the preceding theorem and loss function $L(\xi, d) = (d - \xi)^2$, the MRE estimator is given by*

$$\delta^*(\mathbf{X}) = \frac{\int_{-\infty}^{+\infty} uf(x_1 - u, \ldots, x_n - u)du}{\int_{-\infty}^{+\infty} f(x_1 - u, \ldots, x_n - u)du}. \tag{4.10}$$

Proof *Consider a trivial equivariant estimator for location transformations $\delta_0(\mathbf{X}) = X_n$. To use the previous theorem in the construction of MRE, we need to compute $E(X_n|\mathbf{y})$.*

We can now consider the change of variables $y_i = x_i - x_n$ $(i = 1, \ldots, n-1)$; $y_n = x_n$. The Jacobin of the transformation is 1. The joint density of $Y's$ is therefore

$$p_Y(y_1, \ldots y_n) = f(y_1 + y_n, \ldots, y_{n-1} + y_n, y_n),$$

and the conditional density of Y_n given $\mathbf{y} = (y_1, \ldots, y_{n-1})$ is

$$\frac{f(y_1 + y_n, \ldots, y_{n-1} + y_n, y_n)}{\int f(y_1 + t, \ldots, y_{n-1} + t, t)dt}$$

It follows that

$$E_0[X_n|\mathbf{y}] = E_0[Y_n|\mathbf{y}] = \frac{\int tf(y_1 + t, \ldots, y_{n-1} + t, t)dt}{\int f(y_1 + t, \ldots, y_{n-1} + t, t)dt}$$

This can be re-expressed in terms of the $x's$ as

$$E[X_n|\mathbf{y}] = \frac{\int tf(x_1 - x_n + t, \ldots, x_{n-1} - x_n + t, t)dt}{\int f(x_1 - x_n + t, \ldots, x_{n-1} - x_n + t, t)dt}.$$

Finally by making change of variables $u = x_n - t$, we have

$$E_0[X_n|\mathbf{y}] = x_n - \frac{\int uf(x_1 - u, \ldots, x_n - u)du}{\int f(x_1 - u, \ldots, x_n - u)du}.$$

This completes the proof.

We shall provide an example with a specific model to illustrate the method of restricted optimization for data with location transformations.

Example 4.8 *Assume that the underlying model of a set of data $X_1, ..., X_n$ is uniform $(\xi - 1/2, \xi + 1/2)$. Suppose that the loss function is*

$$L(\xi, d) = (d - \xi)^2,$$

find the MRE of location transformations for ξ.

We may use the approach of Pitman estimator to find the MRE estimator for ξ under location transformations.
Consider

$$f(x_1, ..., x_n) = 1 \text{ for all } x_i \in (\xi - \frac{1}{2}, \xi + \frac{1}{2}); 0 \text{ otherwise.}$$

Notice that

$$x_i \in (\xi - \frac{1}{2}, \xi + \frac{1}{2}) \text{ for all } i = 1, ..., n. \Rightarrow \xi \in (X_{(n)} - \frac{1}{2}, X_{(1)} + \frac{1}{2}).$$

We have the denominator of the Pitman estimator:

$$\int f(x_1 - \xi, ..., x_n - \xi)d\xi = \int_{X_{(n)} - \frac{1}{2}}^{X_{(1)} + \frac{1}{2}} 1 d\xi = X_{(1)} - X_{(n)} + 1.$$

The numerator of the Pitman estimator reads,

$$\int \xi f(x_1 - \xi, ..., x_n - \xi)d\xi = \int_{X_{(n)} - \frac{1}{2}}^{X_{(1)} + \frac{1}{2}} \xi d\xi$$

$$= \frac{1}{2}(X_{(1)} + X_{(n)})(X_{(1)} - X_{(n)} + 1).$$

Thus, by the Pitman estimator, the MRE estimator is

$$\delta* = \frac{\frac{1}{2}(X_{(1)} + X_{(n)})(X_{(1)} - X_{(n)} + 1)}{X_{(1)} - X_{(n)} + 1} = (X_{(1)} + X_{(n)})/2.$$

The next example discusses restricted optimization in the case where the loss function is the absolute error, and the minimization of the risk is restricted to an equivariant estimator for location transformations.

Example 4.9 *Let X be i.i.d. according to the exponential distribution $E(\theta, 1)$, find the MRE of θ for the absolute error $L = |d - \theta|$.*

Solution: By the totality of equivariant estimator $\delta = \delta_0 - v(\mathbf{y})$, and

$$E_\theta(|\delta - \theta|) = E_0(|\delta|) = E_0(|\delta_0 - v(\mathbf{y})|),$$

we need to minimize the risk with respect to v. Since we are not dealing with a square loss function, the usual optimal solution $v^* = E(\delta_0|Y)$ is not applicable here.

Consider the smallest ordered statistic $\delta_0 = x_{(1)} = y_1$, and denote

$$y_i = x_{(i)} - x_{(1)} \quad i = 2, \ldots, n,$$

we have the joint distribution

$$f_{X_{(1)}, \ldots, X_{(n)}} = n! e^{-\sum_{i=1}^n x_i + n\theta},$$

with $|J| = 1$. The distribution of the transformed variables reads,

$$f_{y_1, \ldots, y_n} = n! e^{-\sum_{i=2}^n y_i - x_{(1)} + n\theta}.$$

Now y_1 and $(y_2, \ldots, y_n)$ are independent,

$$E_0(|X_{(1)} - v(\mathbf{y})||\mathbf{y}) = E(|Y_1 - v|).$$

The restricted optimization problem becomes to optimize the term involving v.

$$E_0(|Y_1 - v|) = \int_0^{+\infty} |y - v| n e^{-ny} dy,$$

if $v < 0$,

$$g'(v) = \int_0^{+\infty} (-1) n e^{-ny} dy = -1 < 0,$$

so $g(v)$ is monotone decreasing,

$$\min_v g(v) = \min_{v>0} g(v).$$

Thus,

$$E_0(|Y_1 - v|) = \int_0^{+\infty} |y - v| n e^{-ny} dy$$
$$= \int_0^v (v - y) n e^{-ny} dy = g(v) + \int_v^{+\infty} (y - v) n e^{-ny} dy$$

Setting $g'(v) = 0$ yields that the optimal solution v^* is the median of $X_{(1)}$,

$$v^* = \frac{\log 2}{n}.$$

Therefore the MRE for location transformations with absolute error loss function reads,

$$\delta^* = X_{(1)} - \frac{log2}{n}.$$

In the discussion above, we focused on MRE estimators for location transformations. We shall now consider the corresponding theory for the scale transformation group. Further materials in this regard can be found in [16].

Example 4.10 *Assume that the underlying model of a set of data* $\mathbf{X} = (X_1, ..., X_n)$ *takes the following form,*

$$\frac{1}{\tau^n} f(\frac{\mathbf{X}}{\tau}) = \frac{1}{\tau^n} f(\frac{x_1}{\tau}, ..., \frac{x_n}{\tau}), \quad \tau > 0 \tag{4.11}$$

where f is known and τ is an unknown scale parameter, this model remains invariant under the transformations $X_i' = bX_i$, $\tau' = b\tau$ for any positive value $b > 0$.

We will consider the following loss function for discussions in the rest of this subsection.

$$L(\tau, d) = \frac{(d - \tau^r)^2}{\tau^{2r}}. \tag{4.12}$$

By the definition, it seems that the set of equivariant estimators for scale transformations is very vague. However, the following theorem shows that, similar to the estimation theory for location transformations, we may construct a general expression for scale equivariant estimators, based on a given scale equivariant estimator.

Theorem 4.9 Totality of scale equivariant estimator: *Assume that the underlying model of a set of data $\mathbf{X}$ is (4.11), and δ_0 is any scale equivariant estimator of τ^r. If*

$$z_i = \frac{x_i}{x_n} \quad (i = 1, ..., n-1) \quad and \quad z_n = \frac{x_n}{|x_n|}. \tag{4.13}$$

Denote $\mathbf{z} = (z_1, ..., z_n)$, an necessary and sufficient condition for δ to be a scale equivariant estimator is that there is a function $w(\mathbf{z})$ such that

$$\delta(\mathbf{X}) = \frac{\delta_0(\mathbf{X})}{w(\mathbf{z})}.$$

With the expression of the domain (totality of scale equivariant estimators) in which the optimization is restricted, we can now discuss the following theorem which identifies the optimal solution (minimizing the risk) with restriction to scale equivariant estimators, the Minimum Risk Equivariant (MRE) estimator.

Theorem 4.10 Pitman Estimator: *Assume that the underlying model for a set of data $\mathbf{X}$ takes the form in (4.11). Consider a vector $\mathbf{z}$ given by (4.13). Further, suppose that the loss function is given by (4.12), and there exists a scale equivariant estimator δ_0 for the parameter of interest τ^r with finite risk. For each $\mathbf{z}$, if there exists a number $w(\mathbf{z}) = w^*(\mathbf{z})$ which minimizes*

$$E_1\{\gamma[\delta_0(\mathbf{X})/w(\mathbf{z})]|\mathbf{z}\}, \tag{4.14}$$

Then an MRE, minimum risk equivariant estimator, can be constructed by

$$\delta^*(\mathbf{X}) = \frac{\delta_0(\mathbf{X})}{w^*(\mathbf{X})}. \tag{4.15}$$

Specifically, the MRE estimator reads,

$$\delta^*(\mathbf{X}) = \frac{\int_0^{+\infty} v^{n+r-1} f(vx_1, \ldots, vx_n) dv}{\int_0^{+\infty} v^{n+2r-1} f(vx_1, \ldots, vx_n) dv}. \tag{4.16}$$

Proof *The proof of the above theorem is similar to the proof of Pitman estimator for location transformations.*

We selected the method of Pitman estimator to illustrate the method of restricted optimization in data analysis, which involves using transformed data for estimation or prediction of an unknown parameter characterizing assumed models. Certainly, there are many interesting results in estimation theory that we are unable to exhaust in this book. Interested readers can find further discussions on this part of materials in the books [81], [80], and [16], among others.

SUMMARY Following the preceding chapter on sensitivity and specificity trade-off, this chapter focuses on another type of trade-off, the bias and variation trade-off. For prediction processes in data science, there are always errors or data fluctuation that make the predicted value consist of reducible errors and irreducible errors. Reducible errors can be ameliorated by enhancing the prediction accuracy with close approximation to the underlying rule governing the data. On the other hand, the irreducible error is the one that can not be improved by refining prediction models because the noise is hidden behind the data. Usually, increasing accuracy (which is equivalent to decreasing prediction bias) is bounded to increase the variation of the predicted value. Towards this end, this chapter uses the concept of uniformly minimum variance unbiased estimator (UMVUE) as an example to delineate methods and procedures to keep the unbiased criterion and to minimize the variation of the estimator. Under the measurement of squared prediction error, UMVUE is one of the restrained optimal solutions.

When the measurement criterion changes for different types of prediction problems with data transformation, the optimal estimator UMVUE needs to be extended to MRE (minimum risk estimator). Certainly, when the risk function is the sum of squared prediction errors, UMVUE is one of MRE. However, when a small amount of bias results in large amount of amelioration in variance, UMVUE is no longer the best MRE. MRE covers more information than UMVUE, especially for discrete type of data in classification problems.

5

Linear Prediction

Linear regression is arguably one of the most commonly *used* and *abused* statistical tools in data science. Its versatility and intuitiveness fits a broad range of applications, from simple linear model such as "when the price increases, the return per item increases" or "the insurance premium decreases as the time spent driving increases". This phenomenon can occur in any setting at any time. Despite this, its ease of use tends to backfire when amateur data analysts mindlessly default to reading the data into software (such as R, Python, or Excel) to obtain a fitted line without checking validity conditions of linear regression. They tend to lack consideration for the rationale behind the methodology. As a result, the inferred conclusion sometimes results in a unreliable statistical prediction.

Traditional textbooks on this topic usually begin with introductory examples, followed by a least squares estimate, inference, and discussion. However, in this chapter, we take a different route. We focus on the validity conditions and precautions with linear regression models in practice, to prevent misuse of the technique from the get-go.

5.1 Pitfalls in linear regressions

The first precaution concerns the intrinsic character of the data. We need to ensure that the data is a random sample representing an intrinsic linear relationship. For instance, one can easily select a set of data in which students with big shoe sizes have high SAT test scores. If we blindly fit the data into a linear model, we may reach an unrealistic conclusion that bigger shoe sizes predict higher SAT test scores (or the other way around). This is obviously a misapplication of the tool.

To avoid making such a fundamental error, two basic conditions must be satisfied before fitting the data into a regression model. The first condition is that the data needs to represent the population of interest. In other words, it should be a random sample from the population of interest. The second condition is that the response variable and the predictor should have some intrinsically linear relationship. For instance, the shoe size of an exam taker has nothing to do with the corresponding SAT score. The body weight of cows has no connection with the body weight of rhinos, as illustrated in Figure 5.1. If there is no intuitive relationship between the input and output variables, it would not make sense to set up a linear model between them, even though the relationship may accidentally appear to be linearly correlated in one dataset due to randomness.

Certainly, in data-oriented analysis, we use exploratory data analysis (such as plotting the data) to seek or approach the true model behind the data. Such a practice should be confined to cases where the data truly represent the variable and no sample selection bias exists. Blindly applying linear regression without proper justification may result in misleading conclusions.

FIGURE 5.1
Non-intrinsic linear relationship between weights on cow and rhino

The second precaution focuses on the sample size and the underlying model of the data. Although the least square estimate of model parameters does not require any distribution of the data, testing on the significance and validity of the fitted model depends on hidden assumptions. For example, the error term of the data follows a normal model with constant variation. Such model assumption is critical in validity analysis of the fitted model. Especially since any software can produce a fitted line out of an input and an output variable, whether that relationship is statistically significant is questionable for many data analyses.

If the data contains too much noise, the effect of the noise overwhelms the effect of the input feature, and the fitted linear model is rendered insignificant. In this case, it is necessary to test whether the noise is too large to claim the existence of a linear relationship. The instruments to perform such tests include the t-test for linear coefficient significance and the F-test for the validity of the model. Note that these tests are built upon the normality assumption with constant variances.

$$y_j = \alpha + \beta x_j + \epsilon_j$$

where $\epsilon_j \sim N(0, \sigma^2)$ for all the observations $j = 1, ..., n$.

When the normality assumption is unfulfilled in the evaluation of model validity, any further evaluation of the fitted model requires a large sample size for asymptotic theory, in order to compensate for the model violation. Thus, when the sample size is not significantly large and the underlying distribution of the error term is not normal, it is inappropriate to use the linear regression model.

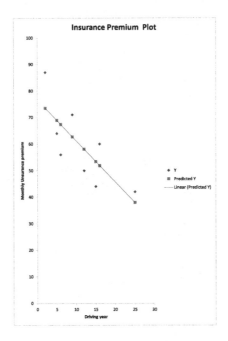

FIGURE 5.2
Bounded driving years vs insurance premium

The third precaution verifies the range of the input features being used in prediction. Recall that the fitted model is built upon training data that is also confined within a range of values of the input predictor. When predicting using the fitted model, if the value of the input variable is beyond the range of the training data, depending on the fitted line, the predicted result may not be meaningful. For instance, in the prediction of insurance premium in Figure 5.2, it does not make sense to extend the linear line into the area of driving age being -1, and claim that the premium is $78.2 per month for a driver at one year before the beginning of one's driving experience, because the intrinsic relationship between the input and the output can not be plausibly extended to that range.

The fourth precaution is the association effect versus the causation effect. It is often confusing and inaccurate to claim that a linear relationship has a causation effect. Linear regression is simply just the fitting of two columns of data points; while the input has an effect on the output, the output may as well influence the input in the linear regression model. In fact, what we can claim in a linear regression analysis is essentially an association effect, not a causation effect.

FIGURE 5.3
Car age vs selling price with outliers

The fifth precaution is the interpretation of the fitted model, especially when there are outliers in the data. As shown in Figure 5.3, on average, the expected price of a brand new car ($X = 0$) is around $20, 574$ with a decrease of around $2338 per year of driving. However, when there are outliers as displayed in Figure 5.3, the fitted line represents

$$selling\ price = 15,813 - 1022 * car\ age,$$

an average decrease of $1022 per driving year and lower estimated price of $15813 for a brand new car.

As shown in the figure, the occurrence of outliers (6.5, 245) flatters the fitted line, and in turn, misrepresents the data pattern for the bulk of the data. Obviously, a value of $24,500 for the selling price of a 6.5-year-old car is an extreme case. In this example, it is more likely that the value of the selling price 245 ($24,500) might be actually 45 ($4500), potentially due to a typo of 2 in front of 45 during the data entry process.

5.2 Model training and prediction

By definition, a simple linear model is represented by a singular input variable and singular output variable that, in tandem, have a potential intrinsically linear relationship. For example, we may choose to model the amount of rainfall an area of fertile land receives alongside the crop yields it produces. This is plausibly linear (within reasonable bounds of rainfall, of course), in conjunction with the idea that more rain may produce a fatter yield. Table 5.1 provides some examples of potentially-linear relationships. For instance, marketing analyst could use sales this month to predict sales next month; the measurement of body coordination is proportionally influenced by the blood alcohol content; and the price of a product may linearly predict the amount that that product sold.

One of the main advantages of linear models is that they are easy to interpret, and the interpretation is immediately intuitive. A raise (or decline) in the input variable produces a proportional result in the output variable. Another feature of linear models is its applicability. Through graphs, the quality of fit in a linear relationship can be visually determined when such a relationship exists.

TABLE 5.1
Intrinsic linear relationships

Input variable	Output variable
Sales this month	Sales next month
Blood alcohol content	Measure of body coordination
Price of a product	Amount of that product sold

In linear regression, the relationship between the input variable and the output variable is usually measured by the sample correlation coefficient, r. If the absolute value of r is close to 1, the linear pattern between the input variable and the output variable is strong. On the other hand, if the value $|r|$ is small, there is essentially not much correlation between the two variables. Under the normality assumption, variables with zero correlation are statistically independent. When the value of Y increases as x increases, the correlation is positive. If the value of Y decreases as the value of x increases, the correlation coefficient is negative. It can be proved that $|r| \leq 1$.

$$r_{xy} = \frac{\sum(x_i y_i) - \frac{1}{n}\sum x \sum y}{\sqrt{\sum x^2 - \frac{1}{n}(\sum x)^2}\sqrt{\sum y^2 - \frac{1}{n}(\sum y)^2}}. \tag{5.1}$$

Equivalently, the sample correlation coefficient is expressed as

$$r_{xy} = \frac{1}{n-1}\frac{\sum(x_i y_i) - \frac{1}{n}\sum x \sum y}{S_x S_y} = \frac{C\hat{O}V(X,Y)}{S_x S_y} \tag{5.2}$$

where $C\hat{O}V(X,Y)$ is the sample covariance between X and Y, S_x and S_y are the sample standard deviation for X, and Y, respectively.

Equation(5.1) and Equation(5.2) are algebraically equivalent. The difference is that Equation (5.2) uses the sample covariance and sample standard deviations that are conventionally used in data analysis. Equation(5.2) also highlights the meaning of correlation coefficient in a way where it measures the relationship between the variation of X and the variation of Y, adjusted by the variations of X and Y.

The following example demonstrates the computation of the sample correlation coefficient r on the basis of a set of bivariate data (X,Y).

Example 5.1 *Assume that we have the following data on the input variable X and output variable Y.*

X:	3	6	12	18	24
Y:	60	95	140	170	185

We can compute the components in the formula (5.1) as follows.

$\sum_i x_i^2 = 1089$	$\sum_i x_i = 63$	$\sum_i x_i y_i = 9930$
$\sum_i y_i^2 = 95350$	$\sum_i y_i = 650$	$n=5$

This results in a sample correlation coefficient of 0.972, indicating a strong linear pattern between the two variables.

Figure 5.4 illustrates different patterns of plots between the input and output variables with their corresponding sample correlation coefficients. When the linear pattern is indiscernible, the sample correlation coefficient is low, at a range from 0 to 0.3. As the linear pattern becomes significant, the absolute value of the corresponding sample correlation coefficient increases. When variable Y increases as X increases, the correlation coefficient is positive; otherwise, it is negative. Figure 5.4 shows that the correlation of sample data is intuitive to understand. The plots of the data cloud fit well with the values of the sample correlation coefficients. The relationship between the input and output variables is easily obtained from a quick glance at the plot of the training data.

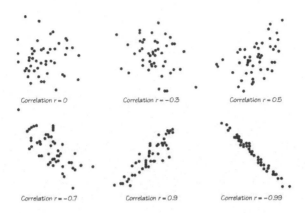

FIGURE 5.4
Data cloud and sample correlation coefficients

5.2.1 Building models with training data

In this subsection, we shall discuss the optimization process for the derivation of the model parameters in a linear model with a set of training data. Recall a general setting in data science where we are interested in obtaining the model behind a set of data by minimizing the expected prediction error.

Step 1. Identifying the shape of the underlying function by optimizing the expected prediction error

When we have a set of training data consisting of a response variable y and a set of predictors $X = (X_1, ..., X_p)^T$, consider the simplicity of a linear function in the optimization process for linear regression. The distance measurement is in terms of the expected prediction error. This can be formulated mathematically as follows.

For a vector $\beta \in R^p$, we have that the function $c(X)$ can be obtained by the typical minimization of prediction error,

$$\hat{c}(X) = \arg\min E_{Y|X}\left([Y - c(X)]^2 \mid X = x\right).$$

Since

$$E((Y - c(X))^2|X) = E((Y - E(Y|X))^2|X) + (E(Y|X) - c(X))^2,$$

the optimal selection of the underlying function is

$$c(X) = E(Y|X).$$

Now, for the data represented by a linear model

$$y = \alpha + \beta X + \epsilon,$$

where ϵ is the term of random error satisfying

$$E(\epsilon) = 0, \quad var(\epsilon) = \sigma^2.$$

This leads to the underlying function

$$c(X) = E(Y \mid X) = \alpha + \beta X.$$

Therefore, when the unknown underlying function $c(X)$ takes on the value $\alpha + \beta X$, the conditional expected value reaches its minimum value, which consequently minimizes the expected prediction errors.

Notice that the above derivation does not require the distribution of the error term. When we search for the optimal solution for the underlying function behind the data, we do not need the distribution pattern of the error term. However, under the assumption that the underlying error term follows a normal model, as discussed in most introductory statistics textbooks, p-values are available to determine whether there is significant data evidence to support the validity of the linear model. In the following sections, we shall address prediction methods using simple linear models with and without normality assumptions in data science, before discussing the implications of multiple linear regression at the end of this chapter.

Step 2. Using the least squares estimation with training data to build a trained model without normality assumptions

Given a set of data consisting of a predictor X and a response Y, to build a simple linear regression model, we need to estimate the coefficients α and β in the model,

$$Y = \alpha + \beta X + \epsilon,$$

where ϵ is the random term of the data. If ϵ follows a normal model, we have existing estimation and testing procedures for the significance on the unknown parameters α and β.

Consider the target function of minimizing the sum of residual errors for a set of training data,

$$L(\alpha, \beta) = \sum_{i=1}^{n}(Y_i - \alpha - \beta X_i)^2. \tag{5.3}$$

Taking derivatives of $L(\alpha, \beta)$ on the variables α and β for a given set of data, gets the optimal solution for Equation(5.3),

$$\hat{\beta} = \frac{\sum_{i=1}^{n} X_i Y_i}{\sum_{i=1}^{n} X_i^2}$$

and

$$\hat{\alpha} = \bar{Y} - \hat{\beta}\bar{X},$$

where

$$\bar{Y} = \frac{1}{n}\sum_{i=1}^{n} Y_i \qquad \bar{X} = \frac{1}{n}\sum_{i=1}^{n} X_i.$$

The above derivation can be alternatively obtained via the following formulation. Consider a case where the target function for optimization is the sample mean squared error $(\hat{MSE})$. On the training set, assume that we have n observation pairs on (x, y) where x is the input variable and y is the output variable. Denote vectors $X = (x_1, ..., x_n)'$ and $Y = (y_1, ..., y_n)'$. Our goal is to minimize

$$\hat{MSE}_{\text{training}} = \frac{1}{n}(Y - \hat{\alpha} - X\hat{\beta})^{\mathsf{T}}(Y - \hat{\alpha} - X\hat{\beta}).$$

After some standard operations as documented in conventional statistics textbooks, the optimal values can be achieved by allowing

$$\hat{\beta} = (X^{\mathsf{T}}X)^{-1}X^{\mathsf{T}}Y, \tag{5.4}$$

and,

$$\hat{\alpha} = \bar{Y} - \hat{\beta}\bar{X}, \tag{5.5}$$

where $\bar{X}$ and $\bar{Y}$ are the sample means of the predictor and the response variable, respectively.

Note that the above derivations do not require any assumption on the normality distribution for the error term ϵ. We shall use an example to illustrate the above discussion.

Example 5.2 *Horizon Properties specializes in custom home re-sales in Phoenix, Arizona. A random sample of 200 records from the custom-home-sale database provides the following information on the size (in hundreds of square feet, rounded to the nearest hundred) and price (in thousands of dollars, rounded to the nearest thousand) of houses in the market.*

Using Equations (5.4) and (5.5), we get an estimated model,

$$y = -110 + 15.89X + \text{error}$$

This is represented in Figure 5.5.

The fitted line can be interpreted as follows. When the house size is 2,000 square feet, the long-run average price in the area is \$207.800. This is because the predicted value $y = -110 + 15.89 \times 20 = 207.8$. Also, each 100-foot increase to the size of a house will increase the long-run average resale value of the house by \$15,890.

Notice that in the interpretation of the regression model, the intercept -110 can not be interpreted directly. Clearly, -110 is a nonsensical value because the value of a house surely cannot be negative. However, this can only occur for a corresponding x-value of 0, implying a house of zero square feet exists, which is also a nonsensical input. This example indicates that a linear regression is only intended to be interpreted within the range of reasonable inputs.

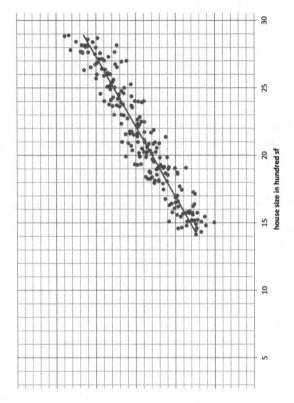

FIGURE 5.5
Horizon Properties, data summary with a linear model

5.2.2 Evaluating trained models without normality

We discuss the sample R^2 and testing MSE (mean squared error) in this subsection, since these two evaluation criteria do not require the normality assumption for the error term in the linear regression model.

1. Sample R^2
    The first and most convenient way is to examine the sample R^2 value in a regression analysis. In a simple linear regression, we can directly calculate the sample R^2 between the input and output variables. In the multiple linear regression case, we have the corresponding multiple correlation matrix for the sample R^2, which essentially serves the same purpose. As it is a good determiner of fit, we may also examine the changes in R^2 when running variable selection algorithms in linear models.
    The sample R^2 is formally defined as the following

$$R^2 = \frac{1}{n-1} \sum_i \left(\frac{x_i - \bar{x}}{s_x} \right) \left(\frac{y_i - \bar{y}}{s_y} \right)$$

where $\bar{x}$ and $\bar{y}$ are the average values of x and y of the training data respectively, and s_x and s_y are the sample standard deviations of the training data.

Taking a closer look at the y terms, we can see that

$$\sum_{i=1}^{n}(y_i - \bar{y})^2 = \sum_{i=1}^{n}(y_i - \hat{y}_i)^2 + \sum_{i=1}^{n}(\hat{y}_i - \bar{y})^2$$

or more succinctly, that

$$SS_{\text{total}} = SS_{\text{residuals}} + SS_{\text{regression}}$$

where SS refers to the sum of squares. That is, the squared sample correlation coefficient R^2 is the fraction of variability in the data explained by the regression model. We can further explore these terms as the following:

SS_{total} Variation in y values

$SS_{\text{residuals}}$ Variation in y from mismatch between model and observed values

$SS_{\text{regression}}$ Variation in y due to changes in x that result in differing output from the model

With the above notations, we have that

$$R^2 = \frac{SS_{\text{regression}}}{SS_{\text{total}}} = \frac{\text{explained variation}}{\text{total variation}}.$$

For a good model, we are able to explain almost all variations, hence the R^2 will be high and approaching 1. If our R^2 value is low (approaching 0), it means that our model does not explain much variation – it is therefore not providing much useful output information for changes in the input variable. This is another indication that the relationship between the input and the output variables is not sensitive enough to claim statistical significance.

2. Sample MSE of the testing data

After building the trained model with training data, it is imperative to assess the validity of the model. Without the assumption of the normality distribution, we do not have p-values to evaluate the significance of the model parameter. Under a data scientist's context, after obtaining the trained model from the training data, we typically evaluate the model using the testing data (usually 25% of the original data).

Given a testing set P, we have that

$$MSE_{\text{test}} = \frac{1}{|P|}\sum_{i \in P}(Y_i - X_i\hat{\beta})^2.$$

In general, theoretically we have the population *expected prediction error*,

$$EPR = E(Y - \hat{g}(X))^2, \tag{5.6}$$

and the sample *test mean squared error* from the testing data,

$$E\hat{P}R = \frac{1}{m}\sum_{i \in P}(Y_i - g(X_i))^2, \tag{5.7}$$

in which m is the size of the testing dataset, i indexes the observation is in the test dataset, and the function g is derived from the training data.

Notice that (5.6) is the population expected prediction error while (5.7) is the sample mean of the individual errors for a set of data. Thus, the reliability of our evaluation on the trained model depends on the sample size in the testing data set. On one hand, if m is large enough, the value of $E\hat{P}R$ is close to the true EPR. On the other hand, if the sample size of the testing data is not large enough, we may underestimate or overestimate the expected prediction error, which consequently leads to misleading conclusion in model selection and consequently misleading prediction.

To better understand how the test expected prediction error (or in this case, the test MSE) relates to the population expected prediction error, we utilize the laws of large numbers.

Theorem 5.1 *Weak Law of Large Numbers.*
For any $\epsilon > 0$, we have

$$\lim_{n \to \infty} P(|\bar{X}_n - \mu| > \epsilon) = 0,$$

for $\mu = E(X_i)$. Equivalently,

$$\bar{X}_n \to^P \mu,$$

as $n \to \infty$.

The weak law of large numbers states that as the number of observations increases, the sample average will be close (in probability) to the expected value. In the setting, the μ is EPR while the sample mean $\bar{X}$ is the sample expected prediction error $E\hat{P}R$. In fact, the relationship between the sample mean and the population mean has a stronger statement:

Theorem 5.2 *Strong Law of Large Numbers.*

$$P(\lim_{n \to \infty} \bar{X}_n = \mu) = 1,$$

for $\mu = E(X_i)$, or equivalently that

$$\bar{X}_n \to^{a.s.} \mu,$$

as $n \to \infty$.

When the sample size is large enough, the average of the observations is almost surely the expected value.

Now, if we let

$$Z_i = (Y_i - \hat{\alpha} - X_i \hat{\beta})^2,$$

we can apply the strong law of large numbers to the sequence $\{Z_i\}$, and obtain

$$\frac{1}{m} \sum_{i=1}^{m} (Y_i - \hat{\alpha} - X_i \hat{\beta})^2$$

$$= \frac{1}{m} \sum_{i=1}^{m} Z_i \to^{a.s.} E(Z)$$

$$= E[(Y - \hat{\alpha} - X_i \hat{\beta})^2].$$

Therefore, when the sample size is sufficiently large, the test MSE almost surely equals the expected prediction error. Stated more directly by combining the training set with the testing set, when the sample sizes in the training set and the test set are large enough, the linear model estimated from the least-squared criterion almost surely has the smallest test MSE. Therefore, it is necessary to have large sample sizes in both the training set and the testing set, when we can not plausibly assume that the error term follows a normal model with a common standard deviation.

Example 5.3 *We use this example to show that for simple linear regression, parameters estimated from the training data do not guarantee the variance decomposition principle.*

Denote the training data by T and the testing data by P, we have

$$\sum_{i \in T} (y_i - \bar{y})^2 = \sum_{i \in T} (y_i - \hat{y}_i)^2 + \sum_{i \in T} (\hat{y}_i - \bar{y})^2 \tag{5.8}$$

but, with probability one, we have

$$\sum_{i \in P} (z_i - \bar{z})^2 \neq \sum_{i \in P} (z_i - \hat{z}_i)^2 + \sum_{i \in P} (\hat{z}_i - \bar{z})^2 \tag{5.9}$$

where $\bar{y}$ and $\bar{z}$ are sample means of the responses in the training set and testing set, respectively. $\hat{y}_i$ and $\hat{z}_i$ are the predicted responses corresponding to the predictor in the training set and testing set, respectively. The response in the training set is denoted as y_i and the response in the testing set is denoted as z_i.

First, we consider the validity of (5.8), notice that when we train the linear model with a set of training data, we have

$$\hat{\beta} = \frac{\sum_{j \in T} (x_j - \bar{x})(y_j - \bar{y})}{\sum_{j \in T} (x_j - \bar{x})^2}$$

and

$$\hat{\alpha} = \bar{y} - \hat{\beta}\bar{x}$$

where $\bar{y}$ and $\bar{x}$ are the sample mean of the responses and the sample mean of the predictor in the training data set. Now,

$$\sum_{i \in T}(y_i - \bar{y})^2$$

$$= \sum_{i \in T}(y_i - \hat{y})^2 + \sum_{i \in T}(\hat{y}_i - \bar{y})^2 + \sum_{i \in T}2(\hat{y}_i - \bar{y})(y_i - \hat{y}_i).$$

Notice that the last term in the expansion has the following property in the training data set.

$$\sum_{i \in T}2(\hat{y}_i - \bar{y})(y_i - \hat{y}_i)$$

$$= 2\hat{\beta}\sum_{i \in T}(x_i - \bar{x})(y_i - \hat{y}_i)$$

$$= 2\hat{\beta}\sum_{i \in T}(x_i - \bar{x})(y_i - \bar{y} - \hat{\beta}(\hat{x}_i - \bar{x}))$$

$$= 2\hat{\beta}\left(\sum_{i \in T}(\hat{x}_i - \bar{x})(y_i - \bar{y}) - \sum_{i \in T}(x_i - \bar{x})^2\frac{\sum_{j \in T}(x_j - \bar{x})(y_j - \bar{y})}{\sum_{j \in T}(x_j - \bar{x})^2}\right)$$

$$= 2\hat{b}(0)$$

$$= 0.$$

This concludes (5.8). Now for (5.9), notice that

$$\sum_{i \in P}(z_i - \bar{z})^2$$

$$= \sum_{i \in P}(z_i - \hat{z})^2 + \sum_{i \in P}(\hat{z}_i - \bar{z})^2 + \sum_{i \in P}2(\hat{z}_i - \bar{z})(z_i - \hat{z}_i).$$

Now, the last term in the expression above becomes,

$$\sum_{i \in P}2(\hat{z}_i - \bar{z})(z_i - \hat{z}_i)$$

$$= 2\sum_{i \in P}(\hat{\alpha} + \hat{\beta}x_i - \bar{z})(z_i - \hat{z}_i)$$

$$= 2\sum_{i \in P}(\bar{y} - \hat{\beta}\bar{x} + \hat{\beta}x_i - \bar{z})(z_i - \bar{y} - \hat{\beta}(\hat{x}_i - \bar{x}))$$

$$= 2\sum_{i \in P}(\bar{y} - \bar{z} + \hat{\beta}(\hat{x}_i - \bar{x}))(z_i - \bar{y}) - \sum_{i \in T}(x_i - \bar{x})^2\frac{\sum_{j \in T}(x_j - \bar{x})(y_j - \bar{y})}{\sum_{j \in T}(x_j - \bar{x})^2}\right)$$

$$\neq 0 (\text{with probability 1}).$$

Thus, Equation (5.9) follows.

The above discussions are grounded on information without the normality assumption for the error term in linear regression analyses. However, when the sample size is not large enough, errors in the asymptotic method are not negligible. This necessitates a discussion on the regression model when normality assumptions are satisfied.

5.2.3   Model significance with normal data

Plugging any two columns of data into a regression software, we can always get a fitted line. Some regression lines do indeed reveal insightful information between the input and the output variables. Others are just insignificant random effect due to variation in the data. When we have large amount of data, the discussion on the sample R^2 and testing MSE in the above subsection can be used to evaluate the significance of the fitted model. However, when the data set is not very large (especially there is no explicit criterion on how large is large enough), it is always helpful to utilize any additional distribution information to facilitate the data analytics process.

We will illustrate this aspect using a hypothetical example of predicting a car resale price based on the age of the car.

Example 5.4 *As shown in the two fitted lines regarding the selling price and the age of the car in Figure 5.3, the top one is for a normal data without outliers while the bottom one corresponds to the occurrence of outliers in the dataset. Although the fitted line corresponding to the dataset without outliers effectively depicts the main pattern of the data cloud, the one with outliers is questionable. Without further information, even with the fitted line, it is debatable to claim the validity of the estimated model. This issue will be further discussed with model significance under normality assumptions below.*

Figure 5.6 shows regression outputs when the underlying distribution of the data is assumed to be normal with a constant variation. As shown in the first part of Figure 5.6, when there are no outliers, the dataset conveys strong linear pattern with $R^2 = 0.978$, and the model is extremely significant with $p - value = 1.35E - 09$. Consider the data variation on the standard error (1.115) for the estimated slope (-23.382), the t-statistic (which is a uniformly most powerful unbiased test), takes the value of -20.978 (estimated regression coefficient adjusted by the sample variation). Thus, the fitted line closely reveals the data pattern between the years of usage and the selling prices of the car. In other words, for the hypothesis

$$H_0 : \beta = 0 \ vs \ H_1 : \beta \neq 0,$$

where $\beta = 0$ indicates that there is no linear relationship between the age of the car and the selling price. $\beta \neq 0$ indicates that the age of the car does affect the selling price of the car. The small p-value indicates data evidence to reject the null hypothesis in favor of the alternative hypothesis.

1. Regression output for car selling price without any outlier.

Regression Statistics	
Multiple R	0.9888279
R Square	0.9777807
Adjusted R Square	0.9755588
Standard Error	5.25746
Observations	12

ANOVA

	df	SS	MS	F	Significance F
Regression	1	12163.62781	12163.63	440.0593	1.34528E-09
Residual	10	276.4088543	27.64089		
Total	11	12440.03667			

	Coefficients	Standard Error	t Stat	P-value	Lower 95%	Upper 95%
Intercept	205.74396	5.33816682	38.54206	3.3E-12	193.8497836	217.63814
Years of car	-23.381625	1.114600159	-20.9776	1.35E-09	-25.8651087	-20.898141

2. Regression output for car selling price with outliers.

Regression Statistics	
Multiple R	0.288086
R Square	0.082993
Adjusted R Square	-0.00037
Standard Error	51.87716
Observations	13

ANOVA

	df	SS	MS	F	Significance F
Regression	1	2679.272	2679.272	0.995553	0.33983196
Residual	11	29603.64	2691.24		
Total	12	32282.91			

	Coefficients	Standard Error	t Stat	P-value	Lower 95%	Upper 95%
Intercept	158.1272	50.65999	3.121344	0.009727	46.6253475	269.62911
Years of car	-10.2281	10.25097	-0.99777	0.339832	-32.790379	12.33408

FIGURE 5.6
Regression outputs on selling prices with outliers

When examining the second part of Figure 5.6, the story deviates from the first portion of the output. With the presence of outliers, the data variation becomes too large and the fitted line essentially becomes insignificant. The sample $R^2 = 0.083$, indicates that there is essentially no linear pattern behind the data.

Notice that the fitted line is moved away from the original one due to the occurrence of outliers. The estimated standard error for the regression coefficient is 10.25, which is almost the same as the absolute value of the regression coefficient (-10.23). Since the data variation is almost as large as the quantity of the estimated model coefficient, the t-statistic is -0.998, with a p-value = 0.3398 suggesting that, with the inclusion of outliers, the data variation has increased to a level that overwhelms the significance of the estimated slope (p=0.3398). Thus, we fail to reject the null hypothesis that $\beta = 0$, which is equivalent to stating that there is no statistical evidence to claim a relationship between the age of a car and its selling price. Under the normality assumptions, the numerical output agrees with the observation on the large variation of the data in the second part of Figure 5.3.

5.2.4 Confidence prediction with trained models

Once the model is trained and validated, the next step is to predict the unknown response based on the explanatory variable x. For instance, when we have the trained model as

$$selling\ price = 20,574 - 2,338\ age\ of\ the\ car,$$

the expected selling price for a car with 6.2 years of usage can be predicted, on average, as

$$selling\ price = 20574 - 2338 * 6.2 = \$6078.40.$$

However, the point estimate is unstable and risky in statistical inference due to randomness in the data. To reach a $(1-\alpha)\%$ confidence level for the predicted value, under the normality assumption, we have the confidence prediction interval,

$$\hat{y} - t_\alpha s_{\hat{y}}, \hat{y} + t_\alpha s_{\hat{y}},$$

where the cut-off value t_α for the Student-t model satisfies,

$$P(|t_{n-2}| < t_\alpha) = 1 - \alpha.$$

The sample standard deviation of $\hat{y}$, $s_{\hat{y}}$ reads

$$s_{\hat{y}} = s_\epsilon \sqrt{1 + \frac{1}{n} + \frac{(x_0 - \bar{x})^2}{SS_{xx}}},$$

where s_ϵ is the sample standard error of residuals, x_0 is the value of the explanatory variable X for which we predict the response, and SS_{xx} is the sum of squares of the explanatory variable X from the training data.

$$SS_{xx} = \sum_i (x_i - \bar{x})^2.$$

5.3 Multiple linear regression

In practice, most linear models will require multiple inputs to capture more nuanced behavior. For instance, in the car resale price example, besides the age of the car, variables affecting the selling price include mileages, the condition of the car, location, special features, shape, and type of the car. Thus, it makes more sense to model the response with a multiple linear model:

$$Y = \alpha + \beta_1 X_1 + ... + \beta_k X_k + \epsilon.$$

Besides inheriting common features in simple linear regression, with more than one predictors in the model, multiple linear regression possesses other discernible properties as discussed below.

Example 5.5 *The following table gives the percentages of concentration of a person, Y, in terms of a set of testing questions, as a response variable with possible association with three predictors including the dosage of a medicine not exceeding the maximum dosage level (X_1), patient age (X_2), and patient stress level (evaluated by a psychologist), X_3.*

ID:	1	2	3	4	5	6	7	8	9	10	11	12
Y:	85	93	79	98	83	66	53	68	72	81	74	87
X_1:	1.7	1.9	1.6	2.0	1.8	1.1	0.9	1.0	1.3	1.5	1.3	1.9
X_2:	23	19	21	22	28	36	34	29	21	32	19	41
X_3:	5.1	4.3	6.2	3.2	3.7	7.2	8.1	7.6	5.3	5.2	6.2	4.1

Task-1) We are interested in knowing whether all the predictors significantly affect the patient concentration level.

Task-2) We are interested in selecting the best model to detect the relationship between the subject concentration level and significant predictors.

```
Call:
lm(formula = concentration ~ dosage + age + stress, data = mydata2)

Residuals:
   Min      1Q    Median    3Q      Max
-5.2335  -2.2658  0.1126  2.2270  5.5940

Coefficients:
            Estimate Std. Error t value Pr(>|t|)
(Intercept) 46.3646  25.8677    1.792   0.1108
dosage      27.7346   9.1358    3.036   0.0162 *
age         -0.2037   0.1635   -1.246   0.2481
stress      -0.7614   2.1805   -0.349   0.7360
---
Signif. codes: 0 '***' 0.001 '**' 0.01 '*' 0.05 '.' 0.1 ' ' 1

Residual standard error: 3.881 on 8 degrees of freedom

Multiple R-squared: 0.9295,    Adjusted R-squared: 0.9031

F-statistic: 35.18 on 3 and 8 DF, p-value: 5.894e-05

> confint(fit, level=0.95)
               2.5 %        97.5 %
(Intercept) -13.2864481  106.0155482

dosage        6.6673741   48.8018131

age          -0.5806767    0.1733577

stress       -5.7897912    4.2668920
```

FIGURE 5.7

Regression of concentration on dosage, age, and stress level

Figure 5.7 contains outcomes of data analysis using multiple linear regression. Among the three predictors, the data only contains significant evidence to claim that the percentage of concentration is linearly affected by the dosage of the medicine. The corresponding p-value is 0.0162, measured by the uniformly most powerful test under the normality assumption of the error term. For each 10% increase in dosage, the corresponding percentage of concentration increases 2.78%, on average, with a 95% confidence interval from 0.67% to 4.88%.

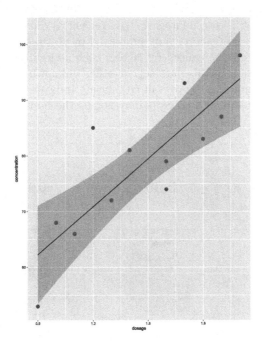

FIGURE 5.8
Dosage-concentration level plot

The overall sample R^2 is at the level of 0.93, suggesting a strong linear pattern of the patient concentration level attributed by the dosage. Figure 5.8 shows the plot of the data between the dosage level and the concentration level.

Although the model between dosage and concentration level is statistically significant, we do know whether this regression model is the best model among all the candidate models involving all three predictors. Note that the total number of candidate linear regression models reads,

$$\binom{p}{0} + \binom{p}{1} + \cdots + \binom{p}{p-1} + \binom{p}{p} = 2^p = 2^3 = 8.$$

When the number of predictors is large, it is practically impossible to examine all the candidate models one-by-one to select the best one. In the sequel, we will discuss multiple correlation matrices among the response and all the predictors, followed by a discussion on the AIC criterion for variable selection in multiple linear regression.

5.3.1 Confounding effects

Following the discussion on the concentration-dosage example, stress is clinically a significant factor related to the concentration level of the patient. However, such a relationship is not supported by this set of data as shown in Figure 5.7. This may be caused by factors such as limited sample size, relatively large variation in this data set, or other unknown reasons. As it turns out, carefully examining the overall correlation indicates that the regression model involving all three predictors may not be the best model for the data.

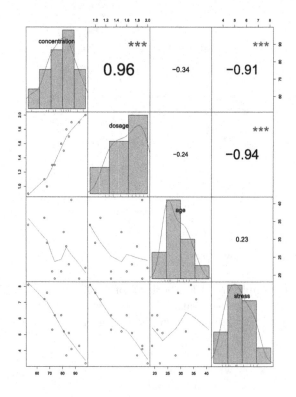

FIGURE 5.9

Correlations on concentration, dosage, age, and stress

Since we are dealing with the response (concentration level), dosage, age, and stress level, the correlation among these variables is a matrix. As shown in Figure 5.9, the concentration level has strong positive correlation with the dosage. Patients taking higher dosage of the medicine tend to have higher level of concentration, on average. The sample correlation coefficient is 0.96 with strong data evidence (extremely small p-value) to reject the null hypothesis H_{10} in favor of the alternative hypothesis H_{11}, where

$$H_{10} : \rho = 0 \quad versus \quad \rho_{concentration-dosage} \neq 0,$$

ρ is the correlation coefficient between the patient concentration percentage and the corresponding dosage.

Although the above analysis coincides with the analysis result in Figure 5.7, further examining the correlation between the concentration level and the stress level reveals the fact that dosage is not the only valuable predictor in our model; stress level is also strongly and negatively associated with the concentration level. Patients with higher stress level tend to have less concentration percentage.

$$\hat{\rho}_1 = -0.91 \qquad \hat{\rho}_1 = -0.94,$$

where ρ_1 is the correlation coefficient between the patient's concentration level and the corresponding stress level. The result of correlation analysis shows that both of the above-mentioned correlation coefficients are significantly different from zero. Since the dosage level of a patient is significantly associated with the stress level, it is naive to assume that the two covariates are independent.

On the other hand, examining the sample correlation of age with the concentration level, dosage, and stress level, we found that the sample correlation coefficients are -0.34, -0.24, and 0.23, respectively. The corresponding p-values are larger than 0.05, indicating that there is no data evidence to claim significant correlation between age and any one of the three variables.

When we fit a multiple linear model without the variable dosage, as shown in Figure 5.10, the stress level feature has a much smaller p-value (9.98e-05) than the significance level of dosage on concentration level (0.0162), implying there is stronger statistical evidence to claim a linear relationship between the concentration level and stress level.

Figure 5.11 shows the plot of the data between the stress level and the concentration level. It is observable that as the stress level increases, the level of concentration decreases. The linear pattern of the data is reflected by the fitted line.

Although the software produces a regression line as shown in Figure 5.12, the variation of the data invalidates the fitted line. It indirectly supports the claim that age is not a significant factor attributable to patient concentration levels.

5.3.2 Information loss and model selection

The prior results indicate that there are 2^p candidate models for a response and p predictors in a multiple linear regression model. Thus, one of the focuses in

```
Call:
lm(formula = concentration ~ age + stress, data = mydata)

Residuals:
   Min     1Q   Median    3Q     Max
-9.207  -2.467   1.958   3.172   4.711

Coefficients:
             Estimate   Std. Error   t value   Pr(>|t|)
(Intercept)  123.1090    7.5847      16.231    5.68e-08 ***
age           -0.2381    0.2256      -1.056    0.319
stress        -6.9624    1.0556      -6.595    9.98e-05 ***
---
Signif. codes:  0 '***' 0.001 '**' 0.01 '*' 0.05 '.' 0.1 ' ' 1

Residual standard error: 5.368 on 9 degrees of freedom
Multiple R-squared:  0.8484,  Adjusted R-squared:  0.8147
F-statistic: 25.17 on 2 and 9 DF,  p-value: 0.0002059

> confint(fit, level=0.95)
               2.5 %         97.5 %
(Intercept) 105.9511341   140.2667866
age          -0.7484385     0.2721478
stress       -9.3504232    -4.5743585
```

FIGURE 5.10
Drug concentration level associated with age and stress

multiple linear regression turns to identify "proper" variables for the training model. That is, we want to pick out the input variables that are going to produce a good model, without bad qualities like noise or over-fitting. This necessitates a discussion on the AIC (Akaike information criterion) and BIC (Bayesian information criterion) for model selection.

When we use data to construct an estimate $\hat{f}(x)$ for the unknown underlying model $f(x)$ via parameter estimation, we need to consider the information loss from using $\hat{f}$ to replace the true model $f(x)$. This is usually measured by the relative entropy of $\hat{f}$ to f in Kullback-Leibler divergence:

$$D(f||\hat{f})_{KL} = \int \log \frac{f(x)}{\hat{f}(x)} dF(x),$$

where $F(x)$ is the cumulative distribution function associated with $f(x)$. It is proven that the estimated model minimizing the Kullback-Leibler divergence $D(f||\hat{f})$ can be obtained via AIC

$$AIC = 2k - 2\log \hat{L}$$

where k is the number of estimated parameters in the model, and $\hat{L}$ is the maximum value of the likelihood function of the model. Specifically, in multiple linear regression with p predictors, AIC reads

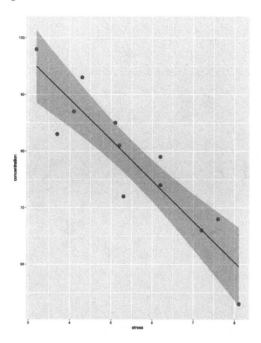

FIGURE 5.11
Drug concentration level associated with stress alone

$$AIC = n + n\log(2\pi) + n\frac{\log(RSS)}{n} + 2(p+1) \qquad (5.10)$$

where RSS is the residual sum of square of the fitted line.

Another commonly used criterion is the BIC (Bayesian information criterion), which reads,

$$BIC = 2\log(n)k - 2\log\hat{L},$$

where n is the number of data points (sample size), k is the number of estimated parameters, and $\hat{L}$ the maximum value of the likelihood function. Correspondingly, for multiple linear regression, the BIC becomes

$$BIC = n + n\log(2\pi) + n\frac{\log(RSS)}{n} + \log(n)(p+1) \qquad (5.11)$$

With the AIC criterion in (5.10) and BIC criteria in (5.11), one can select the model that minimizes the estimated information loss as the best model for the data. However, we can't just test all possibilities of models (that is, every combination of variables). Given p possible input variables, each variable can either be included into the model or not. This means that with p possibilities

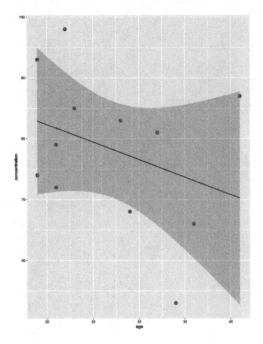

FIGURE 5.12
Drug concentration level associated with age alone

of branching into two, we actually have 2^p possible models. This value quickly becomes intractable. For example, with 10 variables, we have $2^{10} = 1024$ possible models, and with 100 variables, we have $2^{100} = 1.26 \times 10^{30}$ models. Surely, testing each model for viability is not the right way forward. We shall discuss three commonly used variable selection approaches below.

1. Forward Selection

One approach to solve the variable selection problem is known as *forward selection*. This algorithm starts by examining the list of possible explanatory variables and computing a simple regression for each one. The estimated AIC value (or residual sum of squares, or partial F-statistic) is used for the judgment of including an explanatory variable in the final model. Then, we continually add in the next best explanatory variable until no improvement is detected and a final model is achieved.

Computer scientists may recognize this as a *greedy* algorithm. By taking the maximum improvement at each step, we hope to quickly converge on the "best" model.

2. Backward Selection

If we can add variables, so can we subtract. The corresponding *backward selection* algorithm works by running a single massive regression on all possible explanatory variables. With this full-sized regression of all variables, we then examine each one individually to determine which has the smallest estimated AIC value, or partial F-statistic (and hence the largest p-value), and remove this variable. We then continue this process of removing variables until an optimal estimated AIC value or a threshold p-value is reached, leaving us with a final model.

3. Mixed Selection

A blending of the two preceding algorithms is commonly referred to as *mixed selection*. Multiple formulations of this algorithm exist. We use the following as a case to describe the process. For instance, we can use forward selection to add a variable, then run backward selection to retest "prune" any variables, repeating this pair of steps until a satisfied solution is achieved.

```
> step<-stepAIC(fit, direction="both")
Start: AIC=35.68
concentration ~ dosage + age + stress
           Df  Sum of Sq   RSS      AIC
- stress    1    1.837    122.35   33.864
<none>                    120.52   35.683
- age       1   23.376    143.89   35.810
- dosage    1  138.837    259.35   42.879

Step: AIC=33.86
concentration ~ dosage + age
           Df   Sum of Sq   RSS      AIC
<none>                     122.35   33.864
- age       1    23.56     145.92   33.978
+ stress    1     1.84     120.52   35.683
- dosage    1  1390.52    1512.87   62.042

> step<-stepAIC(fit, direction="backward")
Start: AIC=35.68

concentration ~ dosage + age + stress
           Df  Sum of Sq   RSS      AIC
- stress    1    1.837    122.35   33.864
<none>                    120.52   35.683
- age       1   23.376    143.89   35.810
- dosage    1  138.837    259.35   42.879
Step: AIC=33.86
concentration ~ dosage + age
           Df   Sum of Sq   RSS      AIC
<none>                     122.35   33.864
- age       1    23.56     145.92   33.978
- dosage    1  1390.52    1512.87   62.042
```

FIGURE 5.13
AIC criterion and information loss in multiple regression

We shall use the example of the concentration level versus predictors (dosage, age, and stress level) to illustrate the optimizing process for variable selection in multiple linear regression.

Example 5.6 *Refer to Example 5.5, use the AIC criterion to find the best model (the model with the smallest information loss) for the data.*

As shown in Figure 5.13, for the mixed selection approach, the starting model (with all the covariates, dosage, age, and stress level) has AIC 35.68.

$$concentration = \alpha + \beta_1 dosage + \beta_2 age + \beta_3 stress. \qquad (5.12)$$

The AIC with stress deleted from model (5.12) will be 33.864, with age deleted will be 35.81, and with dosage deleted will be 42.879. Since 33.864 < 35.68, in the second step, the variable corresponding to the lowest AIC reduction (stress level) is deleted. For the new model at this step

$$concentration = \alpha + \beta_1 dosage + \beta_2 age. \qquad (5.13)$$

Taking out the variable "age" from the model (5.13) will be 33.978, taking out "dosage" will be 62.042, and adding "stress" back to model (5.13) will be 35.683. Since none of these changes may result in AIC reduction, the best model is model (5.13).

The above is the mixed procedure for the selection of the model that has smallest loss of information. Notice that at the last step, the procedure compares the AIC for adding possible candidate variables in to the model to search for the best model.

Different from the mixed optimization procedure, in the backward search procedure, after the first step of comparing different amounts of AIC changes corresponding to the predictors, the procedure finds only the AIC levels from the removal of the two variables in the model (5.13) without computing the one where the variable "stress" is added back to the model.

Thus, after comparing the AIC values, the model (5.13) is the best linear regression model for the data.

For reference, the R-codes for computations are given in Figure 5.14. It should be noted that in the AIC-optimized model (5.13), the adjusted R^2 is 0.9126, which is larger than 0.9031, the adjusted R^2 for the model (5.12) that uses all the three predictors.

The optimized AIC model (5.13) can be used to predict the percentage concentration level for a patient aged 21 and taking dosage 1.2 ml. As shown in Figure 5.14, the concentration level is predicted as 70.28%, with the 95% confidence prediction interval from 66.19% to 74.36%, when accounting for the randomness from data variation.

5.4 Categorical predictors

Our previous discussion has entirely covered continuous data. We assume that variables like "age" or "weight" can be evaluated to a decimal value. Linear

```
> mydata <- read.csv("/Users/jchen/Desktop/concentration.csv")
> fit<-lm(concentration~dosage+age+stress, data=mydata)
> summary(fit)
> confint(fit, level=0.95)
> fit<-lm(concentration~age+stress, data=mydata2)
> summary(fit)
> confint(fit, level=0.95)

> library(MASS)
> fit<-lm(concentration~dosage+age+stress, data=mydata)
> step<-stepAIC(fit, direction="backward")
> step<-stepAIC(fit, direction="backward")

> fit<-lm(concentration~dosage+age, data=mydata)
> summary(fit)
Call:
lm(formula = concentration ~ dosage + age, data = mydata)
Residuals:
    Min     1Q  Median    3Q    Max
-5.4020 -1.9034 0.0062 2.2340 5.5034
Coefficients:
            Estimate Std. Error  t value   Pr(>|t|)
(Intercept) 37.7029    6.9736     5.407    0.000429 ***
dosage      30.7230    3.0378    10.114    3.26e-06 ***
age         -0.2045    0.1553    -1.317    0.220529
---
Signif. codes:  0 '***' 0.001 '**' 0.01 '*' 0.05 '.' 0.1 ' ' 1

Residual standard error: 3.687 on 9 degrees of freedom
Multiple R-squared: 0.9285,   Adjusted R-squared: 0.9126
F-statistic: 58.4 on 2 and 9 DF,  p-value: 7.006e-06

> predit <- read.csv("/Users/jchen/Desktop/test.csv")
> prediction=predict(fit, newdata=predit, interval='confidence')
> prediction
      fit      lwr        upr
1 70.27688 66.19104  74.36273
```

FIGURE 5.14

R Codes for confidence prediction

regression then works by taking each variable along a continuum and assuming that full interpolation is possible.

However, not all data works cleanly in this form. Some data is *categorical*, meaning that it belongs to certain categories which may have no real relation to each other. For example, the color of car may be important in determining car insurance rates, or the nationality of a car's make in that same model (physicists may complain that color is defined on a wavelength, but this ordering of colors is arbitrary and serves no purpose in this regression).

Other data is just *discrete*, meaning that it will either never come in a continuous form (e.g. star ratings on individual Yelp reviews) or more strongly, perhaps not comprehensible in a continuous form (e.g. take rankings of the best video games – 1.5th place cannot exist). When dealing with well-ordered discrete data, like star ratings, it may make sense to treat the inputs as continuous anyway, but only occurring along a fixed set of values.

However, when dealing with categorical data, we will use *dummy variables* to separate variables. "Dummy variables" are indicator variables taking on the boolean (true/false) value of 0 or 1 based on whether the condition is present. For example, when classifying car insurance rates using colors, we may create dummy variables for each separate color: 1_{red}, 1_{blue}, etc. This notation ($1_{\text{condition}}$) is used to refer to the dummy variable that is 1 when "condition" is true, and 0 otherwise. Of course, we expect only one dummy variable to take on the value of 1 out of the created set. In modern machine learning and computer science, dummy variables are also often referred to as *one-hot encoding* because of this property.

5.5 Outliers and leverage statistics

Any model is subject to the truism "garbage in, garbage out," and linear regressions are no exception. Because the goal is to minimize the sum of square residuals, linear regressions can get greatly influenced to one edge by extreme values, called *outliers*, as discussed in the first section in this chapter. Figure 5.3 shows an example of the impact of outliers that alter the fitted line away from the main-stream of the data, and invalidates the regression output as shown in Figure 5.6.

Outliers are not just points that buck the trend – there can be practical reasons for distinct outliers. For example, in some datasets, a lack of data is simply represented by the maximum or minimum value possible. Regressing on this will easily cause a model that produces nonsensical results. Alternatively, it may represent a literal equipment malfunction when performing a reading. As such, removing these problematic points is critical to deriving a model that is ultimately useful. Many data analysts and modelers spend the majority of their time cleaning and preparing data as input to their models.

In the discussion on outliers, a primary concern is whether outliers are *influential*. Intuitively, this refers to whether outliers are significantly "dragging" the regression away from its ordinarily correct value.

Of course, the concept of *influential point* is not a binary one: we care about the degree to which outliers are influential more than any classification of influential/non-influential points. The study of this metric falls under *leverage statistics*, which is about measuring the distance between an observation and the bulk of the data. For the i-th datum, the leverage score is defined as

$$h_{ii} = \left[X(X^{\mathsf{T}}X)^{-1}X^{\mathsf{T}} \right]_{ii}$$

where X is the design matrix. In other terms, the scores are the diagonal elements of the projection matrix.

If x_i is far away from the average $\bar{x}$, h_{ii} will be large, meaning that the corresponding observation y_i has a large impact on the fitted model. Since the MLE of the linear regression coefficients,

$$\hat{\beta} = (X^{\mathsf{T}}X)^{-1}X^{\mathsf{T}}\mathbf{y},$$

and

$$\hat{\mathbf{y}} = X\hat{\beta} = X(X^{\mathsf{T}}X)^{-1}X^{\mathsf{T}}\mathbf{y},$$

for the ith observation, we have

$$\frac{d(\hat{y}_i)}{dy_i} = \left[X(X^{\mathsf{T}}X)^{-1}X^{\mathsf{T}}\right]_{ii} = h_{ii}.$$

The degree on which the ith response influences the ith predicted value via the observation matrix X.

Proposition 1 In a multiple regression setting, the leverage statistic $0 \leq h_{ii} \leq 1$.

Proof: Since the projection matrix is symmetric and idempotent,

$$(X(X^{\mathsf{T}}X)^{-1}X^{\mathsf{T}})(X(X^{\mathsf{T}}X)^{-1}X^{\mathsf{T}} = X(X^{\mathsf{T}}X)^{-1}X^{\mathsf{T}}).$$

We have

$$h_{ii} = \sum_{t=1}^{n} h_{it}h_{ti} = h_{ii}^2 + \sum_{t \neq i} h_{ti}^2 \geq h_{ii}^2 \tag{5.14}$$

Dividing h_{ii} in both sides of (5.14) gets

$$h_{ii} \leq 1.$$

The first portion of the equation (5.14) implies that

$$h_{ii} \geq 0.$$

Proposition 2 In a multiple regression with n observations and p unknown parameters, the sum of the leverage statistics equal to the dimension of β in the setting

$$\mathbf{y} = X\beta.$$

Proof: Notice that

$$\sum_{i=1}^{n} h_{ii} = \sum_{i=1}^{n} \left[X(X^{\mathsf{T}}X)^{-1}X^{\mathsf{T}}\right]_{ii},$$

we have

$$\sum_{i=1}^{n} h_{ii} = Trace(X(X^{\mathsf{T}}X)^{-1}X^{\mathsf{T}})$$
$$= Trace((X^{\mathsf{T}}X)^{-1}X^{\mathsf{T}}X)$$
$$= Trace(I_p)$$
$$= p.$$

We can take this concept a little further with *Studentized residuals* to test whether the distance is too far away from the bulk of the data.

Given a residual $d_i = y_i - \hat{y}_i$, the variance of d_i can be expressed as:

$$
\begin{aligned}
Var(d_i) &= e_i^\mathsf{T} Var(Y - \hat{Y}) e_i \\
&= e_i^\mathsf{T} Var([I - X(X^\mathsf{T}X)^{-1}X^\mathsf{T}]Y) e_i \\
&= e_i^\mathsf{T}([I - X(X^\mathsf{T}X)^{-1}X^\mathsf{T}]\sigma^2) e_i \\
&= (1 - h_{ii})\sigma^2.
\end{aligned}
$$

Therefore, the Studentized residual is defined as

$$
t_i = \frac{d_i}{\hat{\sigma}\sqrt{1 - h_{ii}}}
$$

This is the residual adjusted for its observation-specific variation.

We shall use an example to explain the formula discussed above.

Example 5.7 *For simplicity, assume that a training data set (toy example) contains*

Y	12	-20	14.3	15.3	15.4
X_1	2.1	2.2	2.3	2.4	2.5
X_2	1.2	1.6	1.7	1.9	1.8

Find and interpret the Studentized residual for the second observation in the training set.

Solution: As shown in Figure 5.15, the leverage statistic for the second observation is

$$
h_{22} = 0.187,
$$

indicating that under the context of the observation matrix X, each unit change of the second response contributes 18.7% to the change of the predicted value of y_2.

The absolute value of the residual of the second response is 17.452. With the residual standard error $\hat{\sigma} = 17.12$, we have

$$
Studentized\, t = \frac{17.452}{(17.12 * \sqrt{(1 - 0.187)})} = 1.13
$$

Thus, at 0.05 significance level, since $t_{0.975,3} = 3.18$ we do not have data evidence to claim that the second observation is significantly beyond the bulk of the data, even though the second response looks well beyond the rest of the response. This is partly due to the relatively small number of observations in the regression.

SUMMARY The method of linear regression prediction is probably one of the most common data science techniques. It is often misused, especially

```
> x<-matrix(c(2.1,1.2,2.2,1.6,2.3,1.7,2.4,1.9,2.5,1.8), 5, 2, byrow=TRUE)
> c=t(x)%*%x
> d=solve(c)
> h=x%*%d%*%t(x)
> h
          [,1]        [,2]        [,3]        [,4]        [,5]
[1,]  0.88237876 0.1155433 0.05645468 -0.2385496 0.1741930
[2,]  0.11554333 0.1870702 0.20082531  0.2338377 0.2090784
[3,]  0.05645468 0.2008253 0.22122392  0.2829465 0.2206971
[4,] -0.23854956 0.2338377 0.28294653  0.4731482 0.2400714
[5,]  0.17419298 0.2090784 0.22069712  0.2400714 0.2361790
> y<-matrix(c(12,-20,14.3, 15.3, 15.4), 5, 1, byrow=TRUE)
> fit<-lm(y~x[,1]+x[,2])
> summary(fit)

Call:
lm(formula = y ~ x[, 1] + x[, 2])

Residuals:
     1      2      3      4      5
 5.717 -17.452 10.030  9.428 -7.723

Coefficients:
             Estimate  Std. Error  t value  Pr(>|t|)
(Intercept) -183.85     172.93      -1.063   0.399
x[, 1]       120.35     113.04       1.065   0.399
x[, 2]       -52.16      66.15      -0.789   0.513

Residual standard error: 17.12 on 2 degrees of freedom

Multiple R-squared: 0.3801,    Adjusted R-squared: -0.2398
F-statistic: 0.6132 on 2 and 2 DF,  p-value: 0.6199
```

FIGURE 5.15
Leverage statistic and outliers

among non-statisticians in a way of coding with input for the purpose of getting an output, without caring about legitimate assumptions and conditions for the regression method. This chapter addresses problems that are commonly abused or overlooked in regression analysis.

Starting with precautions that using linear regression without proper justification may result in misleading conclusions, we discuss the connection between the known and the unknown in identifying the model governing the data. Linear regression is the first step to bridging the known (predictors) and the unknown (response variables) with interpretable functions. In fact, it is the optimal solution when the conditional expectation is indeed a linear combination of the predictors.

We also discuss technical details on training models and testing the trained model with and without normality assumption, a commonly overlooked issue in data science.

Analyzing confounding effects and AIC-BIC criteria for the selection of predictors in a multiple linear model, we use examples to delineate the interpretation, validation, and confidence prediction on linear model. A specific section pertaining to cross-validation on a linear model can be found in Chapter 2.

Probably one of the thorny issues on linear regression is the occurrence of outliers that may alter the bulk of data pattern in regression analysis. At the end of the chapter, we address the issue with theoretical justification and hypothetical data to elucidate a outliers detection method.

6

Nonlinear Prediction

The preceding chapter focuses on methods of regression for a linear function between the predictors and the response. In reality, there are at least two issues restricting the linearity approach. One is the scenario where the number of the predictors is larger than the number of observations, a common issue in big data analytics. Another restriction on the linearity approach occurs in the case where the underlying function is actually not linear. For instance, when the dosage goes beyond the unknown maximal dosage in clinical trials, the dose-response curve may appear to be an umbrella (or inverted-U) rather than a linear relationship. Also, in determining annual premiums for car insurance, drivers with different driving experiences pay different rates (as discussed in the insurance example in Example 1.3). When more predictors are taken into consideration, the underlying model $f(x)$ may take any shape, depending on the mechanism of the relationship between the response and the predictors. This calls for new methodologies for nonlinear predictions.

In this chapter, we shall focus on methods of shrinkage (including ridge regression and LASSO), high-dimension data reduction, polynomial regression, and regression splines.

6.1 Restricted optimization and shrinkage

Recall that in regression analysis with n observation and p predictors, the variation of the prediction

$$\hat{\mathbf{y}} = X\hat{\beta}$$

depends on the variation of the parameter estimate

$$\hat{\beta} = (X^\mathsf{T}X)^{-1}X^\mathsf{T}\mathbf{y}.$$

One way to improve the accuracy of prediction is to reduce the prediction variation, which subsequently puts a restriction on the value range of the parameter β. A properly restricted optimization procedure may produce a more stable prediction in data analytics. Thus, in addition to the least squares optimization in multiple linear regression,

$$\hat{\beta} = \underset{\beta}{argMin} \sum_{i=1}^{n} \left(y_i - \beta_0 - \sum_{j=1}^{p} x_{ij}\beta_j \right)^2, \qquad (6.1)$$

we consider the following two restricted optimizations in the non-linear prediction process.

6.1.1 Ridge regression

$$\hat{\beta}^{ridge} = \underset{\beta}{argMin} \sum_{i=1}^{n} \left(y_i - \beta_0 - \sum_{j=1}^{p} x_{ij}\beta_j \right)^2 \quad subject\ to\ \sum_{j=1}^{p} \beta_j^2 \le t, \quad (6.2)$$

where the value t is a properly selected constant. It is usually determined by a cross-validation process.

The optimization in (6.2) is equivalent to

$$\hat{\beta}^{ridge} = \underset{\beta}{argmin} \left\{ \sum_{i=1}^{n} (y_i - \beta_0 - \sum_{j=1}^{p} x_{ij}\beta_j)^2 + \lambda \sum_{j=1}^{p} \beta_j^2 \right\}$$

for some constant λ. Denote the residual sum of squares

$$RSS = \sum_{i=1}^{n} \left(y_i - \beta_0 - \sum_{j=1}^{p} \beta_j x_{ij} \right)^2,$$

we have

$$\sum_{i=1}^{n} \left(y_i - \beta_0 - \sum_{j=1}^{p} \beta_j x_{ij} \right)^2 + \lambda \sum_{j=1}^{p} \beta_j^2 = RSS + \lambda \sum_{j=1}^{p} \beta_j^2.$$

Denote the updated residual sum of squares,

$$RSS^*(\lambda) = (\mathbf{y} - X\beta)^T (\mathbf{y} - X\beta) + \lambda\beta^T\beta.$$

Minimizing the RSS^* yields the parameter estimates of ridge regression,

$$\hat{\beta}^{ridge} = (X^T X + \lambda I)^{-1} X^T \mathbf{y}. \qquad (6.3)$$

Although this setting has advantages in reducing the prediction error by restricting the range of the model parameters, it should be mentioned that the parameter estimate obtained from (6.2) is not unbiased.

Theorem 6.1 *Model coefficients estimated from the optimization process in (6.2) are biased when $\lambda \neq 0$.*

$$E(\hat{\beta}_{ridge}) \neq \beta.$$

Proof: Notice that by (6.3),

$$\hat{\beta}_{ridge} = (X^T X + \lambda I)^{-1} X^T \mathbf{y}.$$

Denote $\hat{\beta}_{ls}$ the parameter estimate obtained from the least squares estimation (6.1), we have,

$$E(\hat{\beta}_{ls}) = \beta.$$

Let $R = X^T X$, the expectation of the parameter estimators for the ridge regression reads,

$$
\begin{aligned}
E[\hat{\beta}_{ridge}] &= E\{(X^T X + \lambda I)^{-1}(X^T X)[(X^T X)^{-1} X^T \mathbf{y}]\} \\
&= [R(I + \lambda R^{-1})]^{-1} R E[\hat{\beta}_{ls}] \\
&= [(I + \lambda R^{-1})]^{-1}\beta \\
&= [(I + \lambda R^{-1})]^{-1}\beta.
\end{aligned}
$$

Thus, $E(\hat{\beta}_{ridge}) \neq \beta$ if $\lambda \neq 0$.

Example 6.1 *When $p = 1$ and $\alpha = 0$, the optimization problem in (6.2) becomes*

$$\hat{\beta}^{ridge} = \underset{\beta}{argMin} \left\{ \sum_{i=1}^{n}(y_i - x_i\beta_1)^2 + \lambda\beta_1^2 \right\}.$$

Denote

$$g(\beta_1) = \sum_{i=1}^{n}(y_i - x_i\beta_1)^2 + \lambda\beta_1^2$$

and solve the equation

$$\frac{dg(\beta_1)}{d\beta_1} = 0,$$

gets

$$\hat{\beta}_{1ridge} = \frac{\sum_i x_i y_i}{\sum_i x_i^2 + \lambda}.$$

The value of $\hat{\beta}_{1ridge}$ is controlled by the value of λ. As λ increases, $\hat{\beta}_{1ridge}$ shrinks. When λ approaches ∞, $\hat{\beta}_{1ridge}$ approaches zero. When λ takes the value zero, $\hat{\beta}_{1ridge}$ becomes the regular MLE of β_1,

$$\hat{\beta}_{1MLE} = \frac{\sum_i x_i y_i}{\sum_i x_i^2}.$$

6.1.2 LASSO regression

Similar to the restriction of $\sum_{j=1}^{p} \beta_j^2 \leq t$, another approach to eliminate in-substantial and redundant regression coefficients is to restrict the sum of the absolute values of the regression coefficients

$$\sum_{j=1}^{p} |\beta_j| \leq t. \tag{6.4}$$

This approach is named Least Absolute Shrinkage and Selection Operator (LASSO) regression. When the restricted condition in the optimization process (6.2) is replaced by (6.4), we have

$$\hat{\beta}^{lasso} = \underset{\beta}{argMin} \sum_{i=1}^{n} \left(y_i - \beta_0 - \sum_{j=1}^{p} x_{ij}\beta_j \right)^2 \quad subject\ to\ \sum_{j=1}^{p} |\beta_j| \leq t. \tag{6.5}$$

The optimization (6.5) is mathematically equivalent to

$$\hat{\beta}^{lasso} = \underset{\beta}{argmin} \left\{ \sum_{i=1}^{n} (y_i - \beta_0 - \sum_{j=1}^{p} x_{ij}\beta_j)^2 + \lambda \sum_{j=1}^{p} |\beta_j| \right\}.$$

Both optimization problems (6.5) and (6.2) are conditions to prevent over-fitting the training data while minimizing the training MSE,

$$MSE = \sum_{i=1}^{n} \left(y_i - \beta_0 - \sum_{j=1}^{p} x_{ij}\beta_j \right)^2.$$

It should be noted that parameter estimates minimizing the training MSE do not always minimize the testing MSE. However, when both training data and testing data are available, by controlling the scale of estimated parameters like in the ridge regression or LASSO regression, the restriction on the variation of the estimated parameters contributes toward a relatively more stable test MSE.

Example 6.2 *Consider a case where the number of observations $n = 2$, the number of predictors $p = 2$, and the observation matrix X is diagonal with 1's on the diagonal and 0's on off-the diagonal elements. Also assume that the intercept term is zero.*

What is the estimate of β for the usual multiple linear regression? For ridge regression? And for LASSO?

Solution: The model of interest in this example reads

$$\mathbf{y} = X\beta + \epsilon = \beta_1 + \beta_2 + \epsilon$$

For the usual multiple linear regression, the MLE is

$$\hat{\beta} = (X^T X)^{-1} X^T \mathbf{y} = \mathbf{y}.$$

For the ridge regression, as discussed in (6.3),

$$\hat{\beta} = (X^T X + \lambda I)^{-1} X^T \mathbf{y} = \frac{1}{1+\lambda}\mathbf{y}.$$

The estimate will shrink to zero as λ increases.

For the Lasso regression, since the restriction is

$$|\beta_1| + |\beta_2| \leq t,$$

the corresponding target function reads

$$f(\beta_1, \beta_2) = \sum_{j=1}^{2}(y_j^2 - 2y_j\beta_j + \beta_j^2 + \lambda|\beta_j|). \tag{6.6}$$

Thus, by considering $\beta_j > 0$ and $\beta_j < 0$ for (6.6), gets

$$\hat{\beta}_j^L = \begin{cases} y_j - \lambda/2 & y_j > \lambda/2; \\ y_j + \lambda/2 & y_j < -\lambda/2; \\ 0 & |y_j| \leq \lambda/2. \end{cases}$$

6.2   Model Selection and Regularization

As shown in Example 6.2, ridge regression shrinks redundant parameters to a small value by introducing the parameter λ. On the other hand, LASSO directly shrinks them to zero when the sample condition is not satisfied. In fact, this method is similar to the forward, backward, or step-wise approach in model selection where the mean squared error is used as the target function in optimization. These approaches can be unified as practices in model regularization.

Notice that the three model selection approaches (forward, backward, and step-wise) can be expressed with an indicator function and a value s, as follows.

$$\sum_{j=1}^{p} I(\beta_j \neq 0) \leq s,$$

where the predictor x_j is dropped off of by directly setting $\beta_j = 0$.

Thus, the method optimizing MSE in the step-wise model selection for linear model regularization is,

$$\min_{\beta}\{\sum_{i=1}^{n}\left(y_i - \beta_0 - \sum_{j=1}^{p}\beta_j x_{ij}\right)^2\} \ subject \ to \ \sum_{j=1}^{p}I(\beta_j \neq 0) \leq s. \qquad (6.7)$$

The method of ridge regression, in this setting, becomes,

$$\hat{\beta}^{ridge} = arg\underset{\beta}{Min}\sum_{i=1}^{n}(y_i - \beta_0 - \sum_{j=1}^{p}x_{ij}\beta_j)^2 \ subject \ to \ \sum_{j=1}^{p}\beta_j^2 \leq t.$$

and similarly, the method of LASSO in model regularization:

$$\hat{\beta}^{lasso} = arg\underset{\beta}{Min}\sum_{i=1}^{N}(y_i - \beta_0 - \sum_{j=1}^{p}x_{ij}\beta_j)^2 \ subject \ to \ \sum_{j=1}^{p}|\beta_j| \leq t.$$

The above three optimizations focus on the MSE as the target function. However, the model selection approach directly regularizes the model by removing redundant regression coefficients and keeping the key predictors. The method of ridge regression shrinks the associated values of the parameters toward zero. While the LASSO method controls the range in which the predictor attributes toward the response variable.

6.3   High Dimensional Data

Notice that the existence of $(XX')^{-1}$ in the solution of MLS for $\hat{\beta}$ in the linear regression model depends on the condition $n > p$, the sample size is larger than the number of parameters in the model. Related references can be found, for example, Xie and Chen (1988, [126], 1990, [127]), among others.

Another challenge in data analysis occurs when the dimensions of data are much larger than the sample size. For instance, in micro-array gene analysis, we are usually tasked with making inference for thousands of genes using the information of tens or hundreds of patients. In facial recognition, the dimension associated with pixels on the image resolution is frequently larger than the number of training images available. In general, the issue of high dimensionality happens for any data set in which $n << p$, where n is the number of observations and p is the number of predictors.

6.3.1   Curse of Dimensionality

Recall the least square estimation of the linear model with n observation and p predictors,

$$\hat{\beta} = (X^T X)^{-1} X^T \mathbf{y},$$

when $p > n$, the inverse matrix $(X^T X)^{-1}$ does not exist, which invalidates the least square estimate of the regression parameters. The randomness of the sample covariance matrix makes it impossible to resolve the problem with the skill of generalized inverse matrix.

When $n < p$, the usual least squares estimate is not unique, but this can be fixed by adding a constraint, such as in the ridge regression. However, when $n << p$, even with adjusted methods such as ridge regression or LASSO, adding predictors (noisy features) may deteriorate the model.

The following theorem shows that with an appropriately selected constant λ, the least square estimate of the ridge regression always exists.

Theorem 6.2 *When* $\lambda > 0, (X^T X + \lambda I)^{-1}$ *exists.*

Proof: Since $X^T X$ is non negative definite, there exists an orthogonal matrix A, so that

$$X^T X = A \begin{bmatrix} \lambda_1 & \dots & 0 \\ \vdots & \dots & \vdots \\ 0 & \dots & \lambda_n \end{bmatrix} A^T \text{ with } \lambda_i \geq 0$$

$$X^T X + \lambda I = A \begin{bmatrix} \lambda_1 + \lambda & \dots & 0 \\ \vdots & \dots & \vdots \\ 0 & \dots & \lambda_n + \lambda \end{bmatrix} A^T$$

Thus

$$Det(X^T X + \lambda I) \neq 0.$$

6.3.2   Dimension Reduction by Transformation

Assume that we have a set of data $y_i, X_{i1}, X_{i2}, ..., X_{ip}, i = 1, .., n$, with the dimension $p >> n$ the sample size. If there exists a linear transformation of the p predictors,

$$Z_m = \sum_{j=1}^{p} \phi_{jm} X_j \; for \; m = 1, ..., M,$$

and $M < p$. We may then use the new set of predictors $Z_1, ..., Z_M$ to avoid the issue of high dimensionality.

The original regression problem contains $p+1$ regression parameters. However, the transformed data contains $M+1$ parameters. If the dimensions of the transformed data $M < n$ the sample size, the regression problem is solved.

$$
\begin{aligned}
\sum_{m=1}^{M} \theta_m z_{im} &= \sum_{m=1}^{M} \theta_m \sum_{j=1}^{p} \phi_{jm} x_{ij} \\
&= \sum_{j=1}^{p} \sum_{m=1}^{M} \theta_m \phi_{jm} x_{ij} \\
&= \sum_{j=1}^{p} \beta_j x_{ij}.
\end{aligned}
$$

Example 6.3 *Consider the situation in which we have four observations for a regression of five predictors $(n < p)$:*

$$
y_i, X_{i1}, X_{i2}, ..., X_{i5}, i = 1, .., 4,
$$

If we consider the linear transformation,

$$
Z_1 = X_1 + X_2; Z_2 = X_3 + X_4; Z_3 = X_5,
$$

the coefficient relationship between the response and the original predictors

$$
Y = \beta_0 + \beta_1 X_1 + ... + \beta_5 X_5
$$

becomes

$$
Y = \theta_0 + \theta_1 Z_1 + \theta_2 Z_2 + \theta_3 Z_3,
$$

for the transformed data. In this way, when we estimate $\theta_i; i = 1, 2, 3$, in the setting of $p < n$, we can indirectly estimate $\beta_i, i = 1, 2, ...5$ using the corresponding linear transformations.

As shown in the above example, the key is to find the optimal transformation of $Z_m, m = 1, ..., M$ that reduces the high dimensionality issue in linear regression. To this end, we discuss two approaches, the principal component transformation and the partial least squares regression.

Method-1: Principal component regression This approach identifies M new features via the M principal components of the sample covariance matrix, then fits a least squares estimate of the response on the M features for prediction.

The Principal Components: $V \Lambda V^T$ gives a spectral decomposition of $X^T X$ where

$$
\Lambda_{p \times p} = diag[\lambda_1, ..., \lambda_p] = diag[\delta_1^2, ..., \delta_p^2] = \Delta^2
$$

with

$$\lambda_1 \geq \ldots \geq \lambda_p \geq 0$$

denoting the non-negative eigenvalues (also known as principal values) of the non-negative matrix $X^T X$, in which the columns of V, v_j, denotes the corresponding orthonormal eigenvector.

Under this setting, Xv_j and v_j, respectively, denote the jth principal components direction (or PCA loading) corresponding to the jth largest principal value λ_j for each $j \in \{1, \ldots, p\}$.

The following is a simple example to review the concept of principal component.

Example 6.4 Let $X = \begin{pmatrix} 2 & 1 & -3 \\ 1 & -2 & 6 \end{pmatrix}$, find the principal components of X.

Solution: We start with the eigenvalues and eigenvectors of $X^T X$ as follows.

$$X^T X = \begin{pmatrix} 5 & 0 & 0 \\ 0 & 5 & -15 \\ 0 & -15 & 45 \end{pmatrix}$$

$$= \begin{pmatrix} 0 & 1 & 0 \\ \frac{-1}{\sqrt{10}} & 0 & 3/\sqrt{10} \\ 3/\sqrt{10} & 0 & 1/\sqrt{10} \end{pmatrix} \begin{pmatrix} 50 & 0 & 0 \\ 0 & 5 & 0 \\ 0 & 0 & 0 \end{pmatrix} \begin{pmatrix} 0 & \frac{-1}{\sqrt{10}} & 3/\sqrt{10} \\ 1 & 0 & 0 \\ 0 & 3/\sqrt{10} & 1/\sqrt{10} \end{pmatrix}$$

$$= \begin{pmatrix} 0 & 5 & 0 \\ -50/\sqrt{10} & 0 & 0 \\ 150/\sqrt{10} & 0 & 0 \end{pmatrix} \begin{pmatrix} 0 & -1/\sqrt{10} & 3/\sqrt{10} \\ 1 & 0 & 0 \\ 0 & 3/\sqrt{10} & 1/\sqrt{10} \end{pmatrix}$$

The eigenvalues of $X^T X$ are 50, 5, 0 with the following eigenvectors,

$$(v_1, v_2, v_3) = \begin{pmatrix} 0 & 1 & 0 \\ 1/\sqrt{10} & 0 & 3/\sqrt{10} \\ 3/\sqrt{10} & 0 & 1/\sqrt{10} \end{pmatrix}$$

$$Z_1 = Xv_1 = \begin{pmatrix} 2 & 1 & -3 \\ 1 & -2 & 6 \end{pmatrix} \begin{pmatrix} 0 \\ -1/\sqrt{10} \\ 3/\sqrt{10} \end{pmatrix} = \begin{pmatrix} -\sqrt{10} \\ 3\sqrt{10} \end{pmatrix}$$

$$Z_2 = Xv_2 = \begin{pmatrix} 2 & 1 & -3 \\ 1 & -2 & 6 \end{pmatrix} \begin{pmatrix} 1 \\ 0 \\ 0 \end{pmatrix} = \begin{pmatrix} 2 \\ 1 \end{pmatrix}$$

$$Z_3 = Xv_3 = \begin{pmatrix} 2 & 1 & -3 \\ 1 & -2 & 6 \end{pmatrix} \begin{pmatrix} 0 \\ 3/\sqrt{10} \\ 1/\sqrt{10} \end{pmatrix} = \begin{pmatrix} 0 \\ 0. \end{pmatrix}$$

Interpretation: The first principal component contains the largest variation (information) of the original data after data rotation of the orthogonal matrix. It keeps the first k principal components (the k largest sample information) in the transformed data when performing dimension reduction.

Method-2:   Partial Least Squares Approach The previous approach uses principal components to construct linear transformed data, which partially eases the burden in data analytics with $n < p$ difficulty. Alternatively, we may use the following approach to construct a transformed data set where the dimensionality does not exceed the sample size.

The *partial least squares approach* identifies M new features using the following method. After standardizing the p predictors, we can find the first feature Z_1 by setting each coefficient ϕ_{1j} to the coefficient of a simple linear regression of Y onto X_j. Then finding the second feature Z_2 by regressing each variable on Z_1, taking the residuals, and find Z_2 using the orthogonalized data in the same way Z_1 was constructed using the original data. Repeat the construction in the same way to find all Z_M.

In dimension reduction, the principal component regression method (PCR) focuses on the information/variation carried by the input data. However, the partial least squares (PLS) regression focuses more on the correlation between the response and the input. Notice that

$$\hat{\beta} = R_{xy}\frac{S_y}{s_x}$$

due to $(X^TX)^{-1}X^TY$ when $y = \beta_0 + \beta X$ in a simple linear regression setting.

Therefore, when comparing the goals of PCR versus PLS, it becomes obvious that PCR retains the maximal possible amount of information contained in the data, while PLS searches for the maximal percentage of response variation explained by the transformed data.

6.4   Polynomial spline regression

The previous sections discuss the method of regression with coefficient regularization or linear transformation to target the difficulties of large dimensional data. However, when the underlying relationship between the response and the predictors is actually not linear, linear-based regression is inappropriate regardless of regularization techniques. As an extension to linear regression techniques, this section addresses the method of polynomial regression and regression splines.

Consider a degree-d polynomial regression as follows,

$$y_i = \beta_0 + \beta_1 x_i + \beta_2 x_i^2 + \beta_3 x_i^3 + \ldots.. + \beta_d x_i^d + \epsilon_i,$$

for a d-degree polynomial with one-variable x. Besides the polynomial effect, consider the piece-wise constant function such as the example in the insurance premium data in Chapter 1.

The piece-wise constant regression can be formulated as follows.

$$y_i = \beta_0 + \beta_1 C_1(x_i) + \beta_2 C_2(x_i) + \dots + \beta_K C_K(x_i) + \epsilon_i,$$

where

$$C_0(X) = I(X < c_1),$$
$$C_1(X) = I(c_1 < X < c_2),$$
$$C_2(X) = I(c_2 < X < c_3),$$

$$\vdots$$

$$C_{K-1}(X) = I(c_{K-1} < X < c_K),$$
$$C_K(X) = I(c_K \le X),$$

and

$$C_0(X) + C_1(X) + \dots + C_K(X) = 1.$$

Combining the piece-wise constant function and polynomial function, gets the **basic function**,

$$y_i = \beta_0 + \beta_1 b_1(x_i) + \beta_2 b_2(x_i) + \dots + \beta_K b_K(x_i) + \epsilon_i.$$

For example, a piece-wise polynomial function may read,

$$y_i = \begin{cases} \beta_{01} + \beta_{11} x_i + \beta_{21} x_i^2 + \beta_{31} x_i^3 + \epsilon_i & \text{if } x_i \le c; \\ \beta_{02} + \beta_{12} x_i + \beta_{22} x_i^2 + \beta_{32} x_i^3 + \epsilon_i & \text{if } x_i \ge c. \end{cases}$$

Similar to the above example, in general, the points where the coefficients change are called **knots** for basic functions. The concept of *knot* and *spline* are defined as follows.

One technique in fitting a smooth curve to a non-linear relationship between the predictor X and the response Y is the *polynomial spline*.

Consider a set of data (X_i, Y_i), $i = 1, \dots, n$ in the range $X_i \in [a, b]$. For a set of points $\xi_j \in [a, b]$, $j = 1, \dots, m$, the input range $[a, b]$ can be partitioned into

$$(a, b) = \bigcup_{j=1}^{m+1} [\xi_{j-1}, \xi_j],$$

where $\xi_0 = a$, $\xi_{m+1} = b$, and

$$a < \xi_1 < \dots < \xi_m < b.$$

In this setting, $\xi_j \quad j = 1, \dots, m$ are called the knots in the interval $[a, b]$.

A **regression spline** is piece-wise degree-d polynomial, with continuity in the derivative up to degree $d - 1$ at each knot. Delving further, a **cubic spline** is with respect to knots ξ_j, $j = 1, \dots, m$ for the function $g(x)$, is a piece-wise degree-3 polynomial, with continuity in the first and the second derivatives at each knot.

$$g_i(x) = \beta_{1i} + \beta_{2i}x + \beta_{3i}x^2 + \beta 4i x^3, \quad x \in [\xi_i, \xi_{i+1}], \quad i = 0, 1, 2, ..., m,$$

and the smoothness condition

$$g_{i-1}(\xi_i+) = g_i(\xi_i-) \quad g'_{i-1}(\xi_i+) = g'_i(\xi_i-) \quad g''_{i-1}(\xi_i+) = g''_i(\xi_i-).$$

Since each piecewise function $g_i(x)$, $i = 0, 1, ..., m$, contains 4 unknown parameters, $\beta_{1i}, \beta_{2i}, \beta_{3i}, \beta_{4i}$. Without any modification, a cubic spline with m knots has $4m + 4$ parameters to be determined from the training data. Because each interval $[\xi_j, \xi_{j+1}]$, has 3 smoothing conditions, these constraints reduce the total degree of freedom to

$$4m + 4 - 3m = m + 4.$$

A natural cubic spline is a cubic spline that fits a constant in each of the intervals $[a_1, \xi_1]$ and $[\xi_m, b]$. As such, the total number of unknown parameters reduces to $m + 4 - 4 = m$.

Example 6.5 *Consider the case where $m = 2$, namely, we have the partition of the x-range in the following way,*

$$(-\infty, \xi_1], \quad [\xi_1, \xi_2], \quad [\xi_2, \infty),$$

and the corresponding models

$$g_1(x) = \beta_{11} + \beta_{21}x + \beta_{31}x^2 + \beta_{41}x^3, \quad x \in (-\infty, \xi_1]$$
$$g_2(x) = \beta_{12} + \beta_{22}x + \beta_{32}x^2 + \beta_{42}x^3, \quad x \in (\xi_1, \xi_2]$$
$$g_3(x) = \beta_{13} + \beta_{23}x + \beta_{33}x^2 + \beta_{43}x^3, \quad x \in (\xi_2, \infty).$$

Among the coefficients β_{ij}, $i = 1, 2, 3, 4$ and $j = 1, 2, 3$, the following conditions also need to be satisfied in the two knots ξ_1 and ξ_2.

$$\beta_{11} + \beta_{21}\xi_1 + \beta_{31}\xi_1^2 + \beta_{41}\xi_1^3 = \beta_{12} + \beta_{22}\xi_1 + \beta_{32}\xi_1^2 + \beta_{42}\xi_1^3$$
$$\beta_{21} + 2\beta_{31}\xi_1 + 3\beta_{41}\xi_1^2 = \beta_{22} + 2\beta_{32}\xi_1 + 3\beta_{42}\xi_1^2$$
$$2\beta_{31} + 6\beta_{41}\xi_1 = 2\beta_{32} + 6\beta_{42}\xi_1$$
$$\beta_{12} + \beta_{22}\xi_2 + \beta_{32}\xi_2^2 + \beta_{42}\xi_2^3 = \beta_{13} + \beta_{23}\xi_2 + \beta_{33}\xi_2^2 + \beta_{43}\xi_2^3$$
$$\beta_{22} + 2\beta_{32}\xi_2 + 3\beta_{42}\xi_2^2 = \beta_{23} + 2\beta_{33}\xi_2 + 3\beta_{43}\xi_2^2$$
$$2\beta_{32} + 6\beta_{42}\xi_2 = 2\beta_{33} + 6\beta_{43}\xi_2$$

Thus, the number of free variables of β_{ij} $i = 1, 2, 3, 4$, $j = 1, 2, 3$ becomes $4 \times 3 - 2 \times 3 = 6$, which is $m + 4$ when $m = 2$.

In a regression spline, if the set of knots $\{\xi_i, i = 1, ..., m\}$ is taken to be the observed data $\{x_i, i = 1, ..., n\}$, and $m = n$, the cubic spline becomes a smooth curve that passes every point of the observation $\{(x_i, y_i) \quad i = 1, ..., n\}$.

A cubic spline with conditions

$$g^{(2)}(a) = g^{(3)}(a) = 0, \qquad g^{(2)}(b) = g^{(3)}(b) = 0$$

is called a natural cubic spline. A natural cubic spline fits a constant line beyond the range of the knots $[\xi_1, \xi_m]$ (in the beginning and the end of the spline),

$$\beta_{31} = \beta_{41} = 0 \qquad \beta_{3,m+1} = \beta_{4,m+1} = 0.$$

This drops the number of free coefficients to $m + 4 - 4 = m$.

Example 6.6 *When $m = 2$, a natural cubic spline fits*

$$g_1(x) = \beta_{11} + \beta_{21}x, \quad x \in (-\infty, \xi_1]$$
$$g_2(x) = \beta_{12} + \beta_{22}x + \beta_{32}x^2 + \beta_{42}x^3 \quad x \in [\xi_1, \xi_2]$$
$$g_3(x) = \beta_{13} + \beta_{23}x \quad x \in [\xi_2, \infty)$$

In conjunction with the smoothness conditions, the total number of free coefficients drops to $4 \times 3 - 2 \times 3 - 4 = 2$, which is the total number of knots in this example.

Given a set of data, (x_1, y_1), ..., (x_n, y_n), when accounting for the closeness and smoothness of the regression spline, $f(x)$, the penalized residual sum of squares

$$RSS(f, \lambda) = \sum_{i=1}^{n}(y_i - f(x_i))^2 + \lambda \int (f''(t))^2 dt.$$

The first term in RSS measures the closeness of the cubic spline $f(x)$ and the observed data. The second term penalizes curvature in the function, and λ establishes the trade-off between the two, a fixed smoothing parameter usually determined by cross-validation.

When $\lambda = 0$, $RSS(f, \lambda) = 0$, disregarding curvature, the prediction $f(x)$ can be found by interpolating the data to make $RSS = 0$. When $\lambda = \infty$, by using a linear function $f(x) = a + bx$ and $f''(x) = 0$, RSS can reaches its maximum value by using the least squares linear fit.

The following theorem shows that the smoothest function interpolating a set of data is the natural cubic spline.

Theorem 6.3 *Given a set of data (x_i, y_i), $i = 1, ..., n$, $x_i \in (a, b)$, among all functions that interpolate all the data points, the natural cubic spline is the smoothest curve connecter the points when the smoothness is measured by*

$$\mu(f) = \int_a^b (f''(x))^2 dx.$$

Proof: Let $g(x)$ be a natural cubic spline in (a, b) with knots x_i, $i = 1, ..., n$. Assume that G is the set of permissible functions that interpolate all the given data points. For any $f(x) \in G$, we want to show that the smoothness measures

$$\mu(f) \geq \mu(g).$$

Consider

$$t(x) = g(x) - f(x),$$

we have

$$f(x) = g(x) - t(x).$$

Taking the second derivatives in both sides of the equation gets

$$f''(x)g''(x) - t''(x)$$

and the smoothness measure of any interpolating function $f(x)$ becomes

$$\mu(f) = \int_a^b f''(x)dx$$
$$= \int_a^b (g''(x) - t''(x))^2 dx$$
$$= \int_a^b (g''(x))^2 dx + \int_a^b (t''(x))^2 dx - 2\int_a^b g''(x)t''(x)dx$$
$$= \mu(g) + \mu(t) - 2\int_a^b g''(x)t''(x)dx.$$

Now,

$$\int_a^b g''(x)t''(x)dx = \int_a^b g''(x)d(t'(x))$$
$$= t'(x)g''(x)|_a^b - \int_a^b t'(x)dg''(x)$$
$$= t'(b)g''(b) - t'(a)g''(a) - \int_a^b t'(x)g^{(3)}(x)dx.$$

Since $g(x)$ is a natural cubic spline, we have

$$g''(a) = g''(b) = 0.$$

Denote $x_0 = a$, $x_{n+1} = b$ and notice that $g^{(3)}(x) = c$ (where c is a constant) for a cubic spline, we have

$$\int_a^b t'(x)g^{(3)}(x)dx = \sum_{i=0}^n \int_{x_i}^{x_{i+1}} g^{(3)}(x)t'(x)dx$$

$$= \sum_{i=0}^n c \int_{x_i}^{x_{i+1}} t'(x)dx$$

$$= c \sum_{i=0}^n (t(x_{i+1}) - t(x_i))$$

$$= c \sum_{i=0}^n (-f(x_{i+1}) + g(x_{i+1}) - g(x_i) + f(x_i))$$

$$= 0.$$

since both $f(x)$ and $g(x)$ interpolate (x_i, y_i), for $i = 1, ..., n$, we have

$$g(x_i) = y_i = f(x_i) \qquad g(x_{i+1}) = y_{i+1} = f(x_{i+1}).$$

Therefore

$$\mu(f) = \mu(g) + \mu(t),$$

which implies that

$$\mu(g) \le \mu(f).$$

SUMMARY Under the circumstances where the linear relationship between the response and the predictors can not be plausibly assumed, one auxiliary approach is non-linear regression. This is particularly the case when we are confronted with high dimensional data, and the sample covariance matrix is stochastically not positively definitive, as highlighted in Xie and Chen (1988 [126]). Given these conditions, what we discussed in the previous chapter for linear regression is not applicable.

We discussed the method of ridge regression (targeting an additional condition to restrict model coefficients with a sum of squared measurements) and LASSO regression (targeting the least absolute shrinkage and selection operator) with focus on the model conditions and interpretations of regression outputs. Continuing on from the linear model relationship in ridge regression and LASSO, we also describe non-linear regression methods such as polynomial spline regression, as well as principal component transformation for dimensional reduction. Further discussion can be found in papers such as [85], [88],

7

Minimum Risk Classification

This chapter discusses methods of classification that use observable information to predict or estimate an unknown category behind the data. Such classification techniques have various applications, for instance, the diagnosis of a disease, the prediction of up or down trends in the stock market, or the occurrence of a criminal activity in a city, just to name a few. Similar to hypothesis testing problems, for each decision made in classification, there is a chance of correctly classifying the observation into (or not into) a given category, as well as the chance of incorrectly classifying the observation into (or not into) another category.

$diagnosis \backslash true$	$leukemia$	$healthy$
$leukemia$	p_1	$1 - p_2$
$healthy$	$1 - p_1$	p_2

where p_1 is the probability of correctly classifying a leukemia patient as positive for leukemia, and p_2 is the probability of correctly diagnosing a healthy patient as being healthy.

Besides the chances of correct and incorrect classification, for each classification criterion g in the set of all possible classification criteria $\mathcal{G}$, the selection of the classification criterion should also consider the loss associated with the classification decision.

$diagnosis \backslash true$	$leukemia$	$healthy$	
$leukemia$	l_{11}	l_{12}	(7.1)
$healthy$	l_{21}	l_{22}	

where l_{ij} is the loss (cost) of classifying a leukemia (healthy) patient as positive for leukemia (healthy). For instance l_{12} is the loss of incorrectly diagnosing a healthy patient as being positive for leukemia.

Certainly, there are many ways of optimization depending on the condition associated with the practical problem. In this chapter, we focus on the minimum risk classification. Namely, we establish the optimal classification criterion as the one that minimizes the overall risk,

$$\hat{C} = \arg \min_{g \in \mathcal{G}} R(g),$$

where

$$R(g) = l_{11}p_1 + l_{21}(1 - p_1) + l_{12}(1 - p_2) + l_{22}p_2.$$

We shall first define a loss function (hence the risk function) associated with the general classification problem. With the proper definition, we may discuss the optimal solution for minimizing the classification risk, followed by the underlying assumptions upon which the classification algorithms are built. Bayesian classification will be the first topic in the list. After addressing the method of Bayesian classification, we shall also discuss the method of logistic regression, which uses odds ratios to predict the likelihood of the occurrence for a dichotomous outcome.

The classification problem in this chapter can be broadly viewed as an optimization problem on estimation or prediction for the true but unknown category of a given observation. In this regard, the inference approach essentially finds the minimum risk prediction (MRP) based on the classification criterion. We will also discuss scenarios where the loss function is changed from the conventional 0-1 loss to any loss function according to different practical situations.

7.1 Zero-one Loss Classification

The two most common applied discriminant functions are the linear discriminant analysis and quadratic discriminant analysis. However, they are valid only under the assumptions that the loss function is 0-1 loss, and the data follow a multivariate normal model. When covariance matrices of the multivariate models are identical among all the categories, the minimum risk prediction (MRP) criterion results in the linear discriminant function. When the covariance matrices are different across different likelihood functions, the MRP criterion leads to the quadratic discriminant function for multivariate normal data.

First, the risk function corresponding to a 0-1 loss in a classification problem can be formulated as follows.

Let G be the set of possible classes in a classification problem with the size $|G| = K$. Consider the 0-1 loss function penalizing prediction errors.

$$L(g, \hat{g}) = \begin{cases} 1, & if\ \hat{g} \neq g \\ 0, & otherwise. \end{cases}$$

The corresponding risk function for a set of observation x, reads,

$$Risk(G(X)) = \sum_{k=1}^{K} L[\mathcal{G}_k, \hat{G}(X)] \Pr(\mathcal{G}_k | X)$$

Under this setting, the minimum risk estimator of the unknown category becomes,

$$\hat{G}(x) = \arg\min_{g \in \mathcal{G}} \sum_{k=1}^{K} L(\mathcal{G}_k, g) \Pr(\mathcal{G}_k | X = x). \tag{7.2}$$

With the $0 - 1$ loss function, the above equation can be simplified to

$$\hat{G}(x) = \arg\min_{g \in \mathcal{G}}[1 - \Pr(g | X = x)],$$

since

$$\sum_{k=1}^{K} \Pr(\mathcal{G}_k | X = x) = 1.$$

$$\hat{G}(x) = \mathcal{G}_k \text{ if } \Pr(\mathcal{G}_k | X = x) = \max_{g \in \mathcal{G}} \Pr(g | X = x). \tag{7.3}$$

This implies that the minimum risk classifier is the one that is associated with the largest posterior probability.

Interpretation: Given a set of data X, heuristically it is natural to classify the observation into the category that has the highest chance to be the true category associated with the data.

7.1.1 Bayesian Discriminant Functions

To illustrate the search for the optimal solution in classification, we consider the case where we only have two classes $Y = 0$ and $Y = 1$. On the basis of the minimum risk criterion discussed above, we classify the observation into the category $Y = 1$ if $P(Y = 1|X) > P(Y = 0|X)$ for each X; and classify the observation into the class $Y = 0$ if $P(Y = 1|X) < P(Y = 0|X)$ for each X.

Since we typically access the likelihood function through a graphical representation (histogram) of the data, it is relevant at this point to discuss a relationship between likelihood functions and posterior probabilities in classification. As stated above, the classification problem is grounded on the comparison of the posterior probabilities.

Consider the distribution of a random variable X with a likelihood function $f(x|\theta)$ with an unknown parameter θ. If we have some prior information on θ, say, the distribution of θ is $g(\theta)$, then after observing x, the "adjusted belief" on the distribution of θ becomes

$$f(\theta|x) = \frac{f(x|\theta)g(\theta)}{f(x)},$$

where $g(\theta)$ is the prior distribution and $f(\theta|x)$ is the posterior distribution. When the parameter θ takes two values $Y = 0$ and $Y = 1$, the following theorem converts the comparison on posterior probabilities to the corresponding likelihood functions.

Theorem 7.1 *For the likelihood functions* $f(x|Y = 0)$ *and* $f(x|Y = 1)$, *we have*

$$\log(\frac{P(Y = 1|X)}{P(Y = 0|X)}) = \log(\frac{f(x|Y = 1)}{f(x|Y = 0)}) + \log(\frac{P(Y = 1)}{P(Y = 0)}).$$

Proof: To understand the above identity, consider any real set $A \subset R^k$ where k is the dimension of the features X. We have

$$P(Y = 1|X \in A)$$
$$= \frac{P(X \in A \cap \{Y = 1\})}{P(X \in A)}$$
$$= \frac{P(X \in A|\{Y = 1\})P(Y = 1)}{P(X \in A)}$$

Now, let $A = (x, x + \delta]^k$

$$\frac{P(X \in A|Y = 1)}{P(X \in A|Y = 0)} = \frac{P(x < X \leq x + \delta|Y = 1)}{P(x < X \leq x + \delta|Y = 0)}$$

$$= \frac{F_1(x + \delta) - F_1(x)}{F_0(x + \delta) - F_0(x)} = \frac{(F_1(x + \delta) - F_1(x))/\delta}{(F_0(x + \delta) - F_0(x))/\delta}$$

Letting $\delta \to 0$ gets

$$\log\left(\frac{P(Y = 1|X = x)}{P(Y = 0|X = x)}\right) = \log\left(\frac{f(x|Y = 1)}{f(x|Y = 0)}\right) + \log\left(\frac{P(Y = 1)}{P(Y = 0)}\right)$$

This proves Theorem-7.1, which converts the comparison on posterior probabilities to the ratio of likelihood functions and prior probabilities associated with the two classes. For illustrative purposes, we now consider a simple example for the case when the observation only has one dimension, $k = 1$, given the above setting.

Example 7.1 *Assume that* $f(x|Y = 1)$ *and* $f(x|Y = 0)$ *are two normal densities with means* μ_1, μ_0 *and common* σ^2, *under the condition of equal priors, we have*

$$\log\left(\frac{f(x|Y = 1)}{f(x|Y = 0)}\right)$$
$$= \log(exp\{\frac{-1}{2\sigma^2}[(x - \mu_1)^2 - (x - \mu_0)^2]\})$$
$$= \frac{1}{\sigma^2}x(\mu_1 - \mu_0) - \frac{1}{2\sigma^2}(\mu_1^2 - \mu_0^2)$$

Thus

$$\log(\frac{P(Y = 1|X = x)}{P(Y = 0|X = x)}) = ax + b$$

for some constants a and b.

Thus, we have a classification criterion (which is the Linear discriminant classification):
When $\delta = ax + b > 0$, classify $Y = 1$; when $ax + b < 0$, classify $Y = 0$.

When we have two or more predictors with a common covariance matrix across all the categories, in which $X \in R^p$ is a vector, the MRE classification criterion becomes

$$g(x) = \log(\frac{P(Y = 1|X = x)}{P(Y = 0|X = x)})$$

where the likelihood function for each category reads

$$f(x) = \frac{1}{(2\pi)^{p/2}|\Sigma|^{1/2}} exp\big(-\frac{1}{2}(x - \mu)^T \Sigma^{-1}(x - \mu)\big).$$

Principle of MRE classification with 0-1 loss: Given patient information, if the posterior probability of class t is the largest among all the posterior probabilities, we classify the patient to class t. Namely,

$$\hat{t} = \arg \max_{k=1,\dots,m} P(G = k|X = x).$$

Notice that

$$\Pr(G = k|X = x) = \frac{f_k(x)\pi_k}{\sum_{j=1}^{K} f_j(x)\pi_j}.$$

When there are k possible outcomes, the minimum risk classifier assigns an observation to the class in which δ_k is the largest

$$\delta_k(x) = x^T \Sigma^{-1} \mu_k - \frac{1}{2}\mu_k^T \Sigma^{-1} \mu_k + \log \pi_k.$$

We describe two classification criteria δ and δ_k above. The difference between δ_k and δ can be summarized as follows.

δ : classification rule for two equal prior classes with one input variable following normal likelihood with equal variances.

δ_k : classification rule for k general prior classes with p input variables following normal likelihood and equal covariance matrices.

In the discussion above, we assume that the standard deviations are identical for different categories. Such an assumption is not always plausible. When the two standard deviations are not the same, the linear discriminant function becomes invalid, and the corresponding minimum risk classifier is actually the quadratic discriminant function for normal data. In this setting, the comparison of the posterior probability motivates us to finding the classification criterion $\delta(x)$.

$$\delta(x) = \log(\frac{P(Y=1|X=x)}{P(Y=0|X=x)})$$

If $\delta(x)$ is positive, we classify the data to $Y = 1$; otherwise, the classification result is $Y = 0$.

We shall use an example to explain the above setting. Consider the classification of two population means when the data follow two normal models with two different standard deviations.

Notice that in this case, for $i = 1, 2$, we have the models

$$f(x|\mu_i, \sigma_i) = \frac{1}{(2\pi)^{1/2}(\sigma_i)^{1/2}} \exp(-\frac{1}{2\sigma_i^2}(x - \mu_i)^2),$$

which leads to the quadratic discriminant function on x as follows.

$$\delta(x) = \frac{1}{2}(\frac{1}{\sigma_0^2} - \frac{1}{\sigma_1^2})x^2 + (\frac{\mu_1}{\sigma_1^2} - \frac{\mu_0}{\sigma_0^2})x + c,$$

where the constant

$$c = \frac{1}{2}(\frac{\mu_0^2}{\sigma_0^2} - \frac{\mu_1^2}{\sigma_1^2}) + log\pi_1 - log\pi_0.$$

When we have more than one predictor (where X is a vector) with unequal covariance matrices from multivariate normal model,

$$\delta_k = \log(\frac{P(Y=1|X=x)}{P(Y=0|X=x)})$$

takes the following form,

$$\delta_k(x) = -\frac{1}{2}(x - \mu_k)^T \Sigma_k^{-1}(x - \mu_k) - \frac{1}{2}\log|\Sigma_k| + \log\pi_k$$
$$= -\frac{1}{2}x^T\Sigma_k^{-1}x + x^T\Sigma_k^{-1}\mu_k - \frac{1}{2}\mu_k^T\Sigma_k^{-1}\mu_k - \frac{1}{2}\log|\Sigma_k| + \log\pi_k$$

To further clarify the above discussion on the linear discriminant function and quadratic discriminant function regarding p input variables for an output of k possible outcomes, we consider the following example of the prediction/classification of diabetes patients.

Example 7.2 *Consider the diagnosis of diabetes patients with four input features: systolic blood pressure, fasting blood glucose level, BMI, and smoking (nicotine intake). The diagnosis outputs include three possible categories: healthy, pre-diabetes, and diabetes. For a new patient, we want to use features of patient information (input variables) to diagnose the clinical outcome (designate the patient into the right category).*

```
> x<-matrix(c(125, 100, 30, 1.1), 4, 1, byrow=TRUE)
> sigma<-matrix(c(1, 0.5, 0, 0, 0.5, 1, 0, 0, 0, 0, 1, 0, 0, 0, 0, 1), 4,
4, byrow=TRUE)
> sigma
       [,1] [,2] [,3] [,4]
[1,] 1.0 0.5   0   0
[2,] 0.5 1.0   0   0
[3,] 0.0 0.0   1   0
[4,] 0.0 0.0   0   1
> m1<-matrix(c(120, 85, 20, 0.04), 4, 1, byrow=TRUE)
> m2<-matrix(c(138, 110, 26, 1.2), 4, 1, byrow=TRUE)
> m3<-matrix(c(150, 160, 31, 1.7), 4, 1, byrow=TRUE)

> d1<-t(x)%*%sigma%*%m1-
0.5*t(m1)%*%sigma%*%m1+log(0.5)
> d1

       [,1]
[1,] 19299.35
> d2<-t(x)%*%sigma%*%m2-
0.5*t(m2)%*%sigma%*%m2+log(0.3)
> d2
       [,1]
[1,] 19304.4
> d3<-t(x)%*%sigma%*%m3-
0.5*t(m3)%*%sigma%*%m3+log(0.2)
> d3
       [,1]
[1,] 16648.32
```

FIGURE 7.1
R-codes for computation on diabetes classification risk.

Under the normality assumption, for each class, the likelihood function takes the form:

$$f(x) = \frac{1}{2\pi^{p/2}|\Sigma|^{1/2}} exp\Big(-\frac{1}{2}(x - \mu)^T \Sigma^{-1}(x - \mu)\Big)$$

Further, assume that the mean vector for the healthy category is μ_1, pre-diabetes μ_2, and diabetes μ_3 as follows.

$$\mu_1 = \begin{bmatrix} 120 \\ 85 \\ 20 \\ 0.04 \end{bmatrix}, \mu_2 = \begin{bmatrix} 138 \\ 110 \\ 26 \\ 1.2 \end{bmatrix}, \quad and \quad \mu_3 = \begin{bmatrix} 150 \\ 160 \\ 31 \\ 1.7 \end{bmatrix},$$

we also assume that the correlation matrix of the four features is the same for healthy, pre-diabetes, and diabetes patients,

$$\Sigma^{-1} = \begin{bmatrix} 1 & 0.5 & 0 & 0 \\ 0.5 & 1 & 0 & 0 \\ 0 & 0 & 1 & 0 \\ 0 & 0 & 0 & 1 \end{bmatrix}.$$

Assume that the disease distribution of the population reads 50% healthy, 30% pre-diabetes, and 20% diabetes. For a patient with $x^T = (125, 100, 30, 1.1)$, we can then use the linear discriminant classification criterion to diagnose whether he is healthy, pre-diabetic, or has diabetes.

Since the correlation matrix is the same across the four features, we have for $k = 1, 2, 3$,

$$\delta_k = x^T \Sigma^{-1} \mu_k - \frac{1}{2} \mu_k^T \Sigma^{-1} \mu_k + log \pi_k.$$

Thus, for any x we have the three classification criteria as follows.

δ_1

$$= \begin{bmatrix} 125 \\ 100 \\ 30 \\ 1.1 \end{bmatrix}^T \begin{bmatrix} 1 & 0.5 & 0 & 0 \\ 0.5 & 1 & 0 & 0 \\ 0 & 0 & 1 & 0 \\ 0 & 0 & 0 & 1 \end{bmatrix} \begin{bmatrix} 120 \\ 85 \\ 20 \\ 0.04 \end{bmatrix} - \frac{1}{2} \begin{bmatrix} 120 \\ 85 \\ 20 \\ 0.04 \end{bmatrix}^T \begin{bmatrix} 1 & 0.5 & 0 & 0 \\ 0.5 & 1 & 0 & 0 \\ 0 & 0 & 1 & 0 \\ 0 & 0 & 0 & 1 \end{bmatrix} \begin{bmatrix} 120 \\ 85 \\ 20 \\ 0.04 \end{bmatrix}$$

$$+ log(0.5)$$

$$= 19299.35.$$

δ_2

$$= \begin{bmatrix} 125 \\ 100 \\ 30 \\ 1.1 \end{bmatrix}^T \begin{bmatrix} 1 & 0.5 & 0 & 0 \\ 0.5 & 1 & 0 & 0 \\ 0 & 0 & 1 & 0 \\ 0 & 0 & 0 & 1 \end{bmatrix} \begin{bmatrix} 138 \\ 110 \\ 26 \\ 1.2 \end{bmatrix} - \frac{1}{2} \begin{bmatrix} 138 \\ 110 \\ 26 \\ 1.2 \end{bmatrix}^T \begin{bmatrix} 1 & 0.5 & 0 & 0 \\ 0.5 & 1 & 0 & 0 \\ 0 & 0 & 1 & 0 \\ 0 & 0 & 0 & 1 \end{bmatrix} \begin{bmatrix} 138 \\ 110 \\ 26 \\ 1.2 \end{bmatrix}$$

$$+ log(0.3)$$

$$= 19304.4.$$

δ_3

$$
= \begin{bmatrix} 125 \\ 100 \\ 30 \\ 1.1 \end{bmatrix}^T \begin{bmatrix} 1 & 0.5 & 0 & 0 \\ 0.5 & 1 & 0 & 0 \\ 0 & 0 & 1 & 0 \\ 0 & 0 & 0 & 1 \end{bmatrix} \begin{bmatrix} 150 \\ 160 \\ 31 \\ 1.7 \end{bmatrix} - \frac{1}{2} \begin{bmatrix} 150 \\ 160 \\ 31 \\ 1.7 \end{bmatrix}^T \begin{bmatrix} 1 & 0.5 & 0 & 0 \\ 0.5 & 1 & 0 & 0 \\ 0 & 0 & 1 & 0 \\ 0 & 0 & 0 & 1 \end{bmatrix} \begin{bmatrix} 150 \\ 160 \\ 31 \\ 1.7 \end{bmatrix}
$$

$$+ log(0.2)$$

$$= 16648.32.$$

Since the second class has the highest δ value, the new patient (who has the four features $(125, 100, 30, 1.1)$ representing systolic blood pressure, fasting glucose level, BMI, and nicotine level) is classified into the pre-diabetes category. This classification minimizes the possible misclassification risk for the given data.

The R-codes for the computation in the above example can be found in Figure 7.1.

We shall use another example to illustrate the quadratic discrimination classification.

Example 7.3 *Outlined below is a hypothetic example of the application of quadratic discriminant function.*
C-reactive protein, ESR (erythrocyte sedimentation rate), and BMI are common factors in the diagnosis of rheumatoid arthritis. Assume that the factors follow normal models with $\mu_{disease} = (2.1, 35, 30)$, $\sigma_{disease} = (0.3, 5, 2.3)$ for C-reactive protein, ESR, and BMI, respectively. Also, assume that correspondingly, $\mu_{healthy} = (0.7, 15, 17)$, $\sigma_{healthy} = (0.2, 3.1, 2)$, the disease rate of rheumatoid arthritis is 30% in the population. We are interested in developing a classification criterion to detect diseased patients.

Assume that the covariance matrices are

$$
\Sigma_{disease} = \begin{bmatrix} 0.09 & 0.45 & 0 \\ 0.45 & 25 & 0 \\ 0 & 0 & 5.29 \end{bmatrix}, \quad \Sigma_{healthy} = \begin{bmatrix} 0.04 & 0 & 0 \\ 0 & 9.61 & 0 \\ 0 & 0 & 4 \end{bmatrix}
$$

In this setting,

$$
\Sigma_{disease}^{-1} = \begin{bmatrix} 12.21 & -0.2198 & 0 \\ -0.2198 & 0.04396 & 0 \\ 0 & 0 & 0.189 \end{bmatrix},
$$

$$
\Sigma_{healthy}^{-1} = \begin{bmatrix} 25 & 0 & 0 \\ 0 & 0.104 & 0 \\ 0 & 0 & 0.25 \end{bmatrix}.
$$

The classification criterion for an observation $x = (x_1, x_2, x_3)$ reads

$$\delta_{healthy} = -\frac{1}{2} \begin{bmatrix} x_1 \\ x_2 \\ x_3 \end{bmatrix}^T \begin{bmatrix} 25 & 0 & 0 \\ 0 & 0.104 & 0 \\ 0 & 0 & 0.25 \end{bmatrix} \begin{bmatrix} x_1 \\ x_2 \\ x_3 \end{bmatrix}$$

$$+ \begin{bmatrix} x_1 \\ x_2 \\ x_3 \end{bmatrix}^T \begin{bmatrix} 25 & 0 & 0 \\ 0 & 0.104 & 0 \\ 0 & 0 & 0.25 \end{bmatrix} \begin{bmatrix} 0.7 \\ 15 \\ 17 \end{bmatrix}$$

$$- \frac{1}{2} \begin{bmatrix} 0.7 \\ 15 \\ 17 \end{bmatrix}^T \begin{bmatrix} 25 & 0 & 0 \\ 0 & 0.104 & 0 \\ 0 & 0 & 0.25 \end{bmatrix} \begin{bmatrix} 0.7 \\ 15 \\ 17 \end{bmatrix} + log(0.7)$$

$$- \frac{1}{2} log(0.04 * 9.61 * 4)$$

$$\delta_{disease} = -\frac{1}{2} \begin{bmatrix} x_1 \\ x_2 \\ x_3 \end{bmatrix}^T \begin{bmatrix} 12.21 & -0.2198 & 0 \\ -0.2198 & 0.04396 & 0 \\ 0 & 0 & 0.189 \end{bmatrix} \begin{bmatrix} x_1 \\ x_2 \\ x_3 \end{bmatrix}$$

$$+ \begin{bmatrix} x_1 \\ x_2 \\ x_3 \end{bmatrix}^T \begin{bmatrix} 12.21 & -0.2198 & 0 \\ -0.2198 & 0.04396 & 0 \\ 0 & 0 & 0.189 \end{bmatrix} \begin{bmatrix} 2.1 \\ 35 \\ 30 \end{bmatrix}$$

$$- \frac{1}{2} \begin{bmatrix} 2.1 \\ 35 \\ 30 \end{bmatrix}^T \begin{bmatrix} 12.21 & -0.2198 & 0 \\ -0.2198 & 0.04396 & 0 \\ 0 & 0 & 0.189 \end{bmatrix} \begin{bmatrix} 2.1 \\ 35 \\ 30 \end{bmatrix} + log(0.3)$$

$$+ \frac{1}{2} log(det(\begin{bmatrix} 12.21 & -0.2198 & 0 \\ -0.2198 & 0.04396 & 0 \\ 0 & 0 & 0.189 \end{bmatrix}))$$

In practice, the parameter vectors μ_1, μ_2, μ_3, and Σ are unknown. They need to be estimated by the training data through cross-validation before the implementation of the classification procedure.

It should be noted that the linear (and quadratic) discriminant function depends on the assumption that the joint distribution of the features follows a multivariate normal model across the three classes. Such assumptions are critical for the validity of the classification procedure. The assumption on equal covariance matrices is also not easy to verify in practice. In what follows, we discuss a different approach in classification involving logistic regression, which does not require the assumption that the data is in the form of a multivariate normal model.

7.1.2   Logistic regression classification

Consider the relationship between the occurrence of an event Y say lung cancer, and potential risk factors such as age, gender, smoking status, hypertension, diabetes, etc. The potential risk factors can be denoted as X_1, ..., X_k. The following equation captures a linear relationship between the logit function of the probability of lung cancer and its predictors,

$$log\frac{p}{1-p} = \alpha + \beta_1 X_1 + ... + \beta_k X_k$$

where $p = P(Y = 1|X)$ is the probability of the occurrence of lung cancer for patients featuring with X.

For a new patient with features x, if we are able to estimate the probability $p = P(Y = 1|X = x)$, according to the MRE criterion for a given data point x, by Equation (7.3), the patient is classified as $Y = 1$ when $P(Y = 1|X) > P(Y = 0|X)$. This is tantamount to $p = P(Y = 1|X) > 0.5$. Thus the classification of claiming $Y = 1$ when the estimated probability $\hat{p} > 0.5$ is actually a minimum risk classifier for the given data.

For convenience, the data frame for a logistic regression model can be briefly outlined as follows.

Example 7.4

ID	gender	age	smoking	diabetes	hypertension	lung − cancer
N0001	M	30	Y	N	Y	Y
N0002	F	43	N	Y	N	N
N0003	M	71	Y	Y	Y	N
N0004	F	63	Y	N	Y	Y
⋮	⋮				⋮	

Model Assumption The assumption of the logistic regression model is that the response follows a binary model with a common chance for the occurrence of the event given a presumed set of features.

The following example explains the interpretation of a logistic regression model.

Example 7.5 *Assume that a fitted model for the relationship between the occurrence of lung cancer, X_1 (dusty work conditions), and X_2 (smoking) is quantified as*

$$log(\frac{p}{1-p}) = -0.4 + 0.3X_1 + 1.6X_2$$

where

$$p = P(lung \quad cancer|X_1, X_2).$$

The estimated model coefficient 1.6 is usually interpreted as: Controlling for work environment conditions, the odds ratio of lung cancer is $e^{1.6} = 4.953$ for *smoking patients. In other words, smoking increases the odds of lung cancer by almost 5 folds when controlling for other risk factors.*

Additionally, for a patient who smokes and works in a dusty environment, the chance of getting lung cancer reads

$$P(Y = 1|X_1 = 1, X_2 = 1) = \frac{exp(-0.4 + 0.3 + 1.6)}{1 + exp(-0.4 + 0.3 + 1.6)} = 0.82 > 0.5$$

Thus, the patient is classified into the group of being more likely to have lung cancer.

```
> backwards=step(glm.fits)
Start:  AIC=60.13
nodes ~ Xray + grade + stage + age + acid

         Df Deviance    AIC
- grade   1   49.097 59.097
- age     1   49.615 59.615
<none>        48.126 60.126
- acid    1   51.572 61.572
- stage   1   52.558 62.558
- Xray    1   55.350 65.350

Step:  AIC=59.1
nodes ~ Xray + stage + age + acid

         Df Deviance    AIC
- age     1   50.660 58.660
<none>        49.097 59.097
- acid    1   52.085 60.085
- stage   1   55.381 63.381
- Xray    1   57.016 65.016

Step:  AIC=58.66
nodes ~ Xray + stage + acid

         Df Deviance    AIC
<none>        50.660 58.660
- acid    1   53.353 59.353
- stage   1   57.059 63.059
- Xray    1   58.613 64.613
```

FIGURE 7.2
Prostate cancer and feature selection in logistic regression

The selection of the factors significantly associated with the response variable is critical in classification when using the logistic regression model. Including insignificant factors in the final model for classification may result in misleading conclusions. One common approach to remedy this is to use the AIC selection criterion. As shown in Figure 7.2, the occurrence of prostate cancer relative to acid, stage, Xray, grade, and age is investigated. When all the predictors are in the model, the AIC is 60.13. With the removal of each predictor, the AIC changes. For instance, when the feature "grade" is removed, the AIC for the new model is 59.097, which is lower than the AIC of the complete model. Since removing grade results in the lowest AIC among all the candidate models, for the first step, grade is removed and the AIC for the updated model is calculated to be 59.1. The process continues until it finds Xray, stage, and acid to be the model with the lowest AIC. At this point, removing any one of the predictors increases the AIC for the updated model, so the process stops.

Similar to the linear discriminant function and the quadratic discriminant function that we discussed in the preceding subsection, a key step in the construction of the classification criterion using logistic regression is the estimation and cross validation of the model coefficients using training data.

7.2 General Loss Functions

The previous section discusses classification approaches based on minimum risk estimation, in which the risk function is based on a 0-1 loss function. However, in practice, we often face scenarios where the loss function is not 0-1. For example, the loss (cost) of misclassifying a healthy person as being a leukemia patient may bear less loss than misclassifying a leukemia patient as being healthy (missing treatment may lead to loss of life). Carelessly applying procedures for 0-1 loss function may result in misleading prediction/estimation conclusions. In this section, we shall discuss methodologies for classification when the loss function is not 0-1. Under this setting, the risk function is correspondingly changed.

We start with a simple example of non 0-1 loss functions.

Example 7.6 *In the classification of three categories with training data X, consider a loss function $L(g_i, G) = i/6$ when $G(X) \neq g_i$, and $L(g_i, G) = 0$ when $G(X) = g_i$ for $i = 1, 2, 3$. Find and interpret the expression for the minimum risk estimator of the true but unknown category.*

Solution: When $G(X) = g_1$, denote the true category is T, we have the risk,

$$R_1(G) = E_X\{0 * P(T = g_1|X) + \frac{2}{6} * P(T = g_2|X) + \frac{3}{6} * P(T = g_3)\}$$

$$= E_X\{t(X) - \frac{1}{6}P(T = g_1|X)\},$$

where

$$t(X) = \frac{1}{3} * P(T = g_1 | X) + \frac{2}{6} * P(T = g_2 | X) + \frac{3}{6} * P(T = g_3)$$

Similarly, for $j = 2, 3$, the risk associated with $G = g_j$ reads

$$R_j(G) = E_X \{ t(X) - \frac{j}{6} P(T = g_j | X) \}$$

Thus, one of the optimal solutions for the MRE of the unknown category is

$$\hat{G} = \arg \max_j \{ \frac{j}{6} P(T = g_j | X) \}$$

This implies that, unlike the optimal classification criterion for 0-1 loss function, the optimal solution for non 0-1 loss is to classify the observation to a category that has the highest posterior probability when weighted by the associated loss.

The interpretation to the general principle is similar to the 0-1 loss scenario in the sense of maximizing the posterior probability with modification of the weight adjustment based on the non 0-1 loss function.

7.3   Local and Universal Optimizations

It should be noted that classification methods discussed in the preceding section (linear discriminant function, quadratic discriminant function, and logistic discriminant approach) are grounded on point-wise optimization for each set of given observations. In another word, the minimum risk estimation

$$\hat{G}(x) = \arg \min_{g \in \mathcal{G}} R(G), \tag{7.4}$$

with the risk function,

$$R(G) = E_X [\sum_{k=1}^{K} L(\mathcal{G}_k, g) \Pr(\mathcal{G}_k | X = x)],$$

is simplified to the sufficient condition of point-wise minimization problem, as discussed in Equation (7.2), for all observations $x \in R^m$,

$$\hat{G}(x) = \arg \min_{g \in \mathcal{G}} \sum_{k=1}^{K} L(\mathcal{G}_k, g) \Pr(\mathcal{G}_k | X = x). \tag{7.5}$$

Clearly, Equation (7.5) is a sufficient but not necessary condition for the optimal risk formulation in Equation (7.4). The condition that Equation (7.5) is valid for all x guarantees the validity of Equation (7.4). Yet, the validity of Equation (7.4) does not imply Equation (7.5). For instance, $X > 0$ implies that $E(X) \geq 0$, however, $E(X) \geq 0$ does not imply $X > 0$.

Therefore, the conditional optimization (since Equation (7.5) conditioned on the given data x) on the inferred risk is different from the universal optimization in (7.4), which does not depend on observation X. Notice that the loss function $L(\mathcal{G}_k, g)$ is not a function of the data x. Equation (7.4) can be simplified as follows:

$$
\begin{aligned}
R(g) &= E_X[\sum_{k=1}^{K} L(\mathcal{G}_k, g) \Pr(\mathcal{G}_k | X = x)] \\
&= \int [\sum_{k=1}^{K} L(\mathcal{G}_k, g) \Pr(\mathcal{G}_k | X = x)] dF(x) \\
&= \sum_{k=1}^{K} L(\mathcal{G}_k, g) \int_X \Pr(\mathcal{G}_k | X = x)] dF(x) \\
&= \sum_{k=1}^{K} L(\mathcal{G}_k, g) \Pr(\mathcal{G}_k).
\end{aligned}
\tag{7.6}
$$

When the loss function is 0-1 loss, the above equation (7.6) becomes

$$
R(g) = 1 - \Pr(\mathcal{G}_{true} = g)
$$

Thus, the MRE becomes

$$
\hat{G} = \arg\max_k \Pr(\mathcal{G}_k = g)
$$

The MRE classification is the one that has the largest chance to be the true but with unknown category under the 0-1 loss function.

7.4 Optimal ROC Classifiers

In the process of training a classifier using the logistic regression model, similar to the sensitivity and specificity problem that we discussed in Chapter 2 and Chapter 3, the decision of classification contains the following possible outcomes:

decision\true category	$\mathcal{G}_k$	not $\mathcal{G}_k$
$\mathcal{G}_k$	sensitivity	false positive
not $\mathcal{G}_k$	false negative	specificity

Unlike the most powerful test where we fix the probability of type-I error and find the procedure that produces the highest power (lowest probability of type-II error), in classification, all four indexes are interpreted and selected according to the most appropriate situation. For instance, when the true but unknown category $\mathcal{G}_k$ is a serious disease, the cost of mis-classifying a diseased patient as a healthy person may result in life-threatening situations. Hence, controlling for the probability of true positive (or false negative) may be of primary concern. On the other hand, when making the decision to operate or not in an emergency room, a false positive error may result in sending a healthy person to an operation room. In that situation, controlling the false positive (hence true negative) may be of higher weight when examining the results of the confusion matrix.

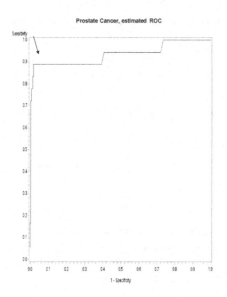

FIGURE 7.3
Prostate cancer, ROC curve, and treatment regimen

One classical approach in logistic regression for balancing the optimal point between the probability of false positive and false negative is to construct an ROC (receiver operating characteristic) curve. The ROC curve usually plots the estimated sensitivity versus 1-specificity (or the sensitivity versus false negative probability). The optimal point (optimal cut-off threshold) is the one corresponding to the point in the ROC curve that has the shortest distance to the golden standard (where sensitivity=1 and specificity=1 for correctly diagnosing all the disease patients and all the healthy patients).

For instance, Figure 7.3 shows an estimated ROC curve in the diagnosis of prostate cancer. When a patient is diagnosed with prostate cancer, an important question emerges in deciding treatment strategy for the patient is whether the cancer cells have spread to the neighboring lymph nodes. The question is so critical during prognosis and treatment that it is customary to operate on the patient (i.e., perform a laparotomy) and remove tissue samples for the sole purpose of examining the nodes for evidence of cancer. However, certain variables can be measured without surgery and are predictive of the nodal involvement. one of such variables is the pathology reading (grade) of a biopsy for the tumor obtained by needle before surgery. Figure 7.3 shows an estimated ROC curve obtained by plotting the sensitivity and 1-specificity at the different grade levels of pathology reading. The point representing the golden standard has 100% success in detecting the spread of prostate cancer and 100% in detecting non-cancer patients, located at (0, 1). Among all the estimated points in the ROC, the one with shortest distance to the golden standard point (0, 1) is the most optimal threshold point. Therefore, the grade level that corresponds to the optimal threshold point is the best cut-off value for a pathology reading in diagnosing of the status of prostate cancer.

As discussed in the prostate cancer example, the optimality criterion now becomes

$$R = \sqrt{(1 - sensitivity)^2 + (1 - sensitivity)^2}$$
$$= \sqrt{(P(false\ positive))^2 + (P(false\ negative))^2},$$

an optimization classifier based on the training data, which is different from the local and universal optimization that we discuss in the previous sections of this chapter.

SUMMARY When predicting a response variable that is categorical (such as "yes" or "no" for a disease, or a different candidate virus in question), the prediction methods discussed in previous chapters, such as linear regression, LASSO, or non-linear regression, can not be applied. This chapter thus focuses on the classification methods and classification criteria that minimize a prediction risk.

For the classification risk function, we start with a zero-one loss function for methods involving logistic regression, which assumes binomial responses. The fitted logistic regression model is then used to compute the predicted probability for classification. Further applications related to logistic regression methods can be found in [15], [29], [30], and [93], among others. After describing the method of logistic regression, we focus on the method of Bayesian discriminant analysis under the normality assumption on the joint distribution of errors. The Bayesian discriminant analysis involves two discriminant functions. We address the difference between linear discriminant function and quadratic discriminant functions in classification. Such description sheds new lights on the hidden condition governing the application of discriminant analysis.

After elucidating methods related to the zero-one loss function, we extend the discussion to applications on general loss functions, which highlights premier conditions before performing data analysis. We address the difference between local and universal optimization, and conclude the chapter with a discussion on the selection of the dose level for optimal ROC classifiers. The ROC criteria essentially cast a new light on the optimization measure that innovates the mean prediction errors discussed in preceding chapters.

8

Support Vectors and Duality Theorem

Support vector machine is a method of classification without making model assumptions on the underlying distribution of the data. For example, consider the classification problem with input vectors X_1, ..., X_n, where $X_i = (x_{i1}, ..., x_{ik})'$ for a response binary output Y, $Y = yes$ or no. The goal of the investigation is to find an effective approach to determine the value of Y on the basis of the information regarding X_1, ..., X_n.

When the input variable X is of high dimension such as in the study of gene expression data or picture recognition in artificial intelligence, the method is especially useful in identifying data points near or beyond the boundary (supporting vectors) to construct optimal classifiers. Theoretically, the optimization process in support vector machine is built upon the duality theorem in linear programming. In this chapter, we start with basic classifiers such as *maximal margin classifier*, *support vector classifier*, and *non-linear boundary classifier* before discussing the mechanism of support vector machine. After describing the methodology, we move deeper into the area by discussing the mechanism in the application of duality theorem. The chapter concludes with a perturbation method in the search for optimal solutions with *duality theorem* in linear programming.

8.1 Maximal Margin Classifier

We start with the following example to illustrate the setting and introduce the concept of maximal margin classifier. The example is about predicting the annual revenue status of a company based on its advertising costs.

Example 8.1 *Consider a set of data for advertising costs (in thousand dollars for TV advertising and Internet advertising) of 20 companies in a soft-drink industry, along with their annual revenue status (positive or negative) in the past year.*

*   **Positive:** (20, 54); (30, 42); (28, 63); (42, 29); (38, 35); (31, 44); (29, 52); (62, 18); (32, 49); (53, 41)*

*   **Negative:** (20, 18); (24, 22); (25, 13); (22, 16); (24, 18); (21, 24); (18, 32); (22, 16); (12, 28); (33, 12)*

If a soft-drink company plans to spend TV advertising budget=30K and Internet advertising budget=40K in the coming year, we want to predict whether the company will have positive or negative annual revenue.

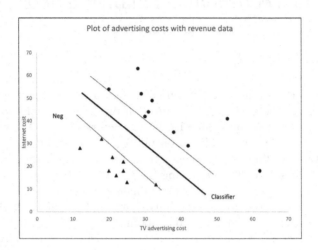

FIGURE 8.1

The hyperplane *classifier* separates companies with positive and negative revenues associated with advertising input of 20 soft-drink companies.

As shown in Figure 8.1, the information contained in the dataset can be depicted in the plot where the TV advertising cost is the x-axis and the Internet advertising cost is the y-axis with the round dot representing positive revenues and triangle dot representing negative revenues. Without any additional model information, it is clear in this example that once the input of the two advertising costs is above the line *positive*, the revenue is positive; while for those below the line *Neg*, the outcome is negative.

Let $Y = 1$ denote the outcome of positive revenue and $Y = -1$ denote the outcome of negative revenue. Denote the boundary for positive revenue be

$$a_1 X_1 + b_1 X_2 + c_1 = 0,$$

and the boundary for negative revenue be

$$a_2 X_1 + b_2 X_2 + c_2 = 0.$$

Intuitively, the *best classifier* is the one that keeps the equal distance between the positive and the negative lines. We call it the maximal margin classifier,

$$a_3 X_1 + b_3 X_2 + c_3 = 0.$$

With this setting, all the data points (X_{i1}, X_{i2}, y_i) satisfy

$$y_i(a_3 X_{i1} + b_3 X_{i2} + c_3) > 0,$$

for $i = 1, ..., n$, because when $a_3 X_1 + b_3 X_2 + c_3 < 0$, the corresponding $y_i < 0$.

As for the company that has $X_1 = 30$, and $X_2 = 40$, the company is expected to have positive revenue because the point $(30, 40)$ locates on the positive side of the classifier.

In the setting of the above example, the key component is the positive line, the negative line, and the classifier. In fact, in the p dimension space, the term we frequently use is the hyperplane.

8.1.1   Hyperplane

In a p-dimensional space, a hyperplane is a flat affine subspace of dimension $p-1$. For instance, the line $x + y = 2$ in $\mathbb{R}^2$ for the xy-plane, where $(x, y) \in \mathbb{R}^2$. Alternatively, the plane $x + 3y + 2z = 10$ in R^3 for $(x, y, z) \in R^3$. Or generally

$$\beta_0 + \beta_1 X_1 + \beta_2 X_2 + ... + \beta_p X_p = 0,$$

for X_i in the R^p space.

Recall that in R^3, a plane with a normal vector (A, B, C) and a point (x_0, y_0, z_0) can be expressed as

$$A(x - x_0) + B(y - y_0) + C(z - z_0) = 0,$$

which is the same as

$$Ax + By + Cz + D = 0,$$

where $D = -Ax_0 - By_0 - Cz_0$.

It is related at this point to review the concept of a *unit normal vector* of a line:

$$\mathbf{n} = \frac{\mathbf{N}}{||\mathbf{N}||} = \frac{(A, B, C)}{\sqrt{A^2 + B^2 + C^2}}.$$

With the concept of *unit normal vector*, the distance between a point (x_1, y_1) and a line featured by $\mathbf{n}$ and (x_0, y_0), becomes

$$\begin{aligned} d &= |\mathbf{v}.\mathbf{n}| \\ &= |(x_1 - x_0, y_1 - y_0).\mathbf{n}| \\ &= \frac{|A(x_1 - x_0) + B(y_1 - y_0) + C(z_1 - z_0)|}{\sqrt{A^2 + B^2 + C^2}}, \end{aligned}$$

which is,

$$d = \frac{Ax_1 + By_1 + Cz_1 + D}{\sqrt{A^2 + B^2 + C^2}}.$$

The above expression can be further simplified as

$$d = \beta_0 + \beta_1 x_1 + \beta_2 y_1 + \beta_3 z_1,$$

where β_i, $i = 1, 2, 3$ satisfies the condition

$$\sum_{i=1}^{3} \beta_i^2 = 1.$$

This directly leads to the maximal margin classifier in the way of finding the classifier that maximizes the distances between the two different types of data points.

8.1.2   Definition of maximal margin classifier

On the basis of the above discussion, the maximal margin classifier can be viewed as a hyperplane that keeps the highest possible values for the data points in each of the two categories. The hyperplane condition necessitates

$$\sum_{j=1}^{p} \beta_j^2 = 1.$$

and the distance condition requires

$$y_i(\beta_0 + \beta_1 x_{i1} + \beta_2 x_{i2} + ... + \beta_p x_{ip}) \geq M.$$

Recall that the distance from a point to a hyperplane

$$d = \frac{|\mathbf{y}.\mathbf{n}|}{||\mathbf{n}||},$$

we have the following definition for the concept of maximal margin classifier.

Definition 8.1 *The maximal margin classifier is a hyperplane that keeps the largest possible distance from the two classes of data. Namely, it is the hyperplane*

$$\beta_0 + \beta_1 x_{i1} + \beta_2 x_{i2} + ... + \beta_p x_{ip} = 0$$

that satisfies the following conditions.

$$\underset{\beta_0,\beta_1,...,\beta_p,M}{maximize\ M}$$

$$subject\ to \sum_{j=1}^{p} \beta_j^2 = 1$$

$$y_i(\beta_0 + \beta_1 x_{i1} + \beta_2 x_{i2} + ... + \beta_p x_{ip}) \geq M\ \forall i = 1, ..., n.$$

The maximal margin classifier for a set of completely separable data can be found by iteratively substituting the current line with an updated line until no further substitution is available. For convenience, we consider the simplest case where $p = 1$. Starting from any point, we can intuitively move the cutoff point x until the two groups are distinctive. And the maximal margin classifier is the middle value between the x value for positive outcome and the x value for negative outcome. Algebraically, to maximize the term

$$D(\beta, \beta_0) = \sum_{i \in \mathcal{M}} y_i(x_i^T \beta + \beta_0),$$

where $\mathcal{M}$ indexes the set of misclassified points, we have

$$\partial \frac{D(\beta, \beta_0)}{\partial \beta} = \sum_{i \in \mathcal{M}} y_i x_i$$

$$\partial \frac{D(\beta, \beta_0)}{\partial \beta_0} = \sum_{i \in \mathcal{M}} y_i.$$

The final solution of the maximal margin classifier can be found by recursively updating

$$\frac{\beta}{\beta_0} \leftarrow \frac{\beta}{\beta_0} + \rho \frac{\beta}{\beta_0},$$

where

$$\rho = \frac{x_i - x_{i0}}{x_{i0}}.$$

The following example shows how to use R to find the maximal margin classifier for a set of separable data. The problem is about the prediction of post-thrombotic syndrome related to the percentage of thrombolysis and the time after the first minor stroke symptoms of stoke patients. More background information on thrombolysis and post-thrombotic syndrome can be found in, for example, Chen and Comerota (2012 [24]).

Example 8.2 *The following is a set of hypothetic data on post-thrombotic syndrome for 10 stroke patients in terms of remaining percentage after thrombolysis X_1 and the time after the onset of the first minor stroke symptoms in minutes X_2.*

Positive post-thrombotic syndrome: (20, 54); (30, 42); (28, 63); (42, 29); (38, 35)

Negative post-thrombotic syndrome: (20, 18); (24, 22); (25, 13); (22, 16); (24, 18)

We are interested in finding the support vectors, maximal margin hyperplane, and two margins to predict post-thrombotic syndrome with the information on percentage remaining after thrombolysis procedures and time passed after the onset of the first minor stroke.

As shown in Figure 8.2, the three support vectors are (42, 29), (30, 42) for patients with positive post thrombotic syndrome, and (24, 22) for negative post-thrombotic syndrome. The line separating the two areas in the diagram is the maximal margin classifier, the margin for positive post-thrombotic syndrome is the line determined by the two support vectors (42, 29), and (30,

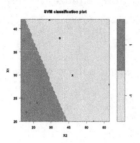

```
x=matrix(c(20, 30, 28, 42, 38, 20, 24, 25, 22, 24, 54, 42, 63, 29, 35, 18, 22, 13, 16,
18), 10, 2)
y=rep(c(-1, 1), c(5, 5))
par(mar=c(1, 1, 1, 1))
plot(x, col=y+3, pch=19)
dat=data.frame(x, y=as.factor(y))
library(e1071)
svmfit=svm(y~., data=dat, kernel="linear", cost=10, scale=FALSE)
dev.new(width=5, height=4)
plot(svmfit, dat)
```

FIGURE 8.2
The hyperplane classifier predicts patients with positive post-thrombotic syndrome based on the percentage of thrombolysis and the time after the first onset of minor stroke symptoms.

42), while the margin for negative post-thrombotic syndrome is the line determined by the slope of the positive post-thrombotic syndrome and passing the support vector (24, 22).

8.2   Support Vector Classifiers

The previous section discusses the method of *maximal margin classifier* in the classification of two categories based on the feature information X, under the *assumption* that the data can be completely separated into two distinct clusters. However, in practice, we are often confronted with situations where the two groups of data are not completely separable, as shown in Figure 8.3. In

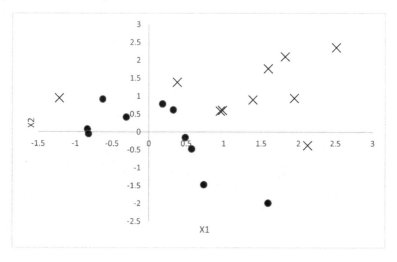

FIGURE 8.3
When data are not linearly separable between the two classes, *maximal margin classifier* does not work

this case, we may either change the linear classifier to a non-linear classifier, or consider an optimization in which some vectors are allowed to be misclassified. The vectors in the margin and in the area of misclassification are the *support vector* in the classification process. This naturally extends the *maximal margin classifier* to the method of *support vector classifier*.

Definition 8.2 *For datasets that are not completely separable, the classifier that optimizes the following target function with the conditions for a given cost C is defined as a support vector classifier.*

$$\underset{\beta_0,\beta_1,\ldots,\beta_p,\epsilon_1,\ldots,\epsilon_n,M}{maximize} \quad M$$

$$subject\ to \sum_{j=1}^{p} \beta_j^2 = 1$$

$$y_i(\beta_0 + \beta_1 x_{i1} + \beta_2 x_{i2} + \ldots + \beta_p x_{ip}) \geq M(1 - \epsilon_i)$$

$$\epsilon_i \geq 0, \quad \sum_{i=1}^{n} \epsilon_i \leq C,$$

In the scenario where the feature vector X has large dimension, the optimization process in the above definition can be simplified by the *duality theorem*, which states that every problem of the form,

$$\text{maximize} \quad \sum_{j=1}^{n} c_j x_j$$

$$\text{subject to} \quad \sum_{j=1}^{n} a_{ij} x_j \leq b_i \quad i = 1, 2, ..., m$$

$$x_j \geq 0 \quad j = 1, 2, ..., n,$$

has a dual:

$$\text{minimize} \quad \sum_{i=1}^{m} b_i y_i$$

$$\text{subject to} \quad \sum_{i=1}^{m} a_{ij} y_i \geq c_j \quad j = 1, 2, ..., n$$

$$y_i \geq 0 \quad i = 1, 2, ..., m.$$

Notice that the original problem is to optimize over R^p space while the dual problem optimizes over the R^1 space. When the dimension p is large, such as in the facial recognition in AI, the target dimension is changed dramatically. We use the following example to illustrate the duality theorem.

Example 8.3 *Consider the question to maximize the profit with fixed resources,*

$$\text{maximize } c_1 x_1 + c_2 x_2 + c_3 x_3$$
$$\text{subject to } a_{11} x_1 + a_{12} x_2 + a_{13} x_3 \leq b_1$$
$$a_{21} x_1 + a_{22} x_2 + a_{23} x_3 \leq b_2$$
$$x_1, x_2, x_3 \geq 0,$$

where c_j=profit per unit of product j produced;
b_i=units of raw material i on hand;
a_{ij}=units raw material i required to produce 1 unit of product j.

The problem can be viewed from the angle of cost as follows. If we save one unit of product j, then we free up:
- a_{1j} units of raw material 1 and
- a_{2j} units of raw material 2.

Selling these unused raw materials at the price of y_1 and y_2 dollars/unit, respectively, yields $a_{1j} y_1 + a_{2j} y_2$ dollars, which is the corresponding cost.

Assume that we are only interested whether the cost exceeds lost profit on each product j:

$$a_{1j} y_1 + a_{2j} y_2 \geq c_j, \quad j = 1, 2, 3.$$

Producing as much product as possible to gain max profit is the same as efficiently minimizing the cost with certain input constraints

$$\text{minimize} \quad b_1 y_1 + b_2 y_2$$
$$\text{subject to} \quad a_{11} y_1 + a_{21} y_2 \geq c_1$$
$$a_{12} y_1 + a_{22} y_2 \geq c_2$$
$$a_{13} y_1 + a_{23} y_2 \geq c_3$$
$$y_1, \; y_2 \geq 0.$$

With the duality theorem, the support vector classifier can be formulated as follows.

$$\min_{\beta, \beta_0} \{ \frac{1}{2} ||\beta||^2 + C \sum_{i=1}^{N} \xi_i \}$$

$$\text{subject to} \quad \xi_i \geq 0, \; y_i(x_i^T \beta + \beta_0) \geq 1 - \xi_i \; \forall i,$$

$$L_P = \frac{1}{2} ||\beta||^2 + C \sum_{i=1}^{N} \xi_i - \sum_{i=1}^{N} \alpha_i [y_i(x_i^T \beta + \beta_0) - (1 - \xi_i)] - \sum_{i=1}^{N} \mu_i \xi_i$$

$$\beta = \sum_{i=1}^{N} \alpha_i y_i x_i,$$

$$0 = \sum_{i=1}^{N} \alpha_i y_i,$$

$$\alpha_i = C - \mu_i, \; \forall i,$$

In this setting, observations that lie directly on the margin, or on the wrong side of the margin for their class, are defined as *support vectors*. These observations affect the construction of the support vector classifier, while the rest of the data do not contribute to the classifier.

In the above formulation, there is a principle on the trade-off between bias and variance in support vector machine. When the value of C is large, there is a high tolerance for observations being on the wrong side of the margin, therefore, the margin will consequently be large. More support vectors which leads to lower variance and higher bias. On the other hand, when the value of C decreases, the tolerance for observations being on the wrong side of the margin decreases, and consequently the margin shrinks. This results in less observations violating the margin and consequently less support vectors. With less support vectors, the support vector classifier has higher variance and lower bias.

We now apply the above techniques to analyze the example in Figure 8.3

Example 8.4 *Since the two data groups are not linearly separable, there is no solution for the setting of maximal margin classifier. We seek the support vector classifier with the cost being set to 10.*

$$min_{\beta,\beta_0}\frac{1}{2}||\beta||^2 + C\sum_{i=1}^{N}\xi_i$$

$$subject\ to\quad \xi_i \geq 0,\ y_i(x_i^T\beta + \beta_0) \geq 1 - \xi_i\ \forall i,$$

The 7 cross points in Figure 8.4 are the support vectors.

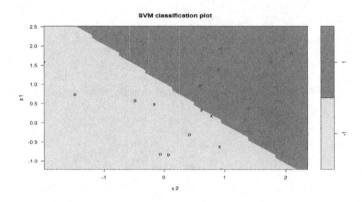

```
library(e1071)
set.seed(1)
x=matrix(rnorm(20*2), ncol=2)
y=c(rep(-1, 10), rep(1, 10))
x[y==1,]=x[y==1,]+1
plot(x, col=(3-y))
dat=data.frame(x=x, y=as.factor(y))
svmfit=svm(y~., data=dat, kernel="linear", cost=10,
scale=FALSE)
plot(svmfit, dat)
```

FIGURE 8.4
When data are not completely separable between the two classes, support
vector classifier works with a cost

8.3   Support Vector Machine

The previous section raises the point that when the maximal margin classifier does not exist, we have to use the support vector classifier with a given cost for misclassified observations or observations on the boundary (support vectors). Following this way of thinking, since the support vector classifier uses a linear boundary to classify the observations, it is possible that the linear classifier determined by the support vectors, may not be able to adequately catch the expected result in classification. Under this scenario, to lower the variation and bias at the cost of straight-forward interpretation, we use a non-linear boundary classifier determined by the support vectors. Since the final result depends on the support vectors, the method of *Support Vector Machine* goes in the way that the classification criteria (output) are directly influenced by the support vectors.

The support vector machine with a quadratic kernel function can be mathematically formulated as follows.

$$\underset{\beta_0,\beta_{11},\beta_{12}\ldots,\beta_{p1},\beta_{p2},\epsilon_1,\ldots,\epsilon_n,M}{\text{maximize}} M$$

$$\text{subject to } y_i\left(\beta_0 + \sum_{j=1}^{p}\beta_{j1}x_{ij} + \sum_{j=1}^{p}\beta_{j2}x_{ij}^2\right) \geq M(1-\epsilon_i)$$

$$\epsilon_i \geq 0, \quad \sum_{i=1}^{n}\epsilon_i \leq C, \quad \sum_{j=1}^{p}\sum_{k=1}^{2}\beta_{jk}^2 = 1.$$

Other common choices of kernel functions in *support vector machine* include

*d*th- Degree polynomial: $K(x,x') = (1 + \langle x,x'\rangle)^d$,

Radial basis: $K(x,x') = \exp(-\gamma||x-x'||^2)$,

Neural network: $K(x,x') = \tanh(\mathcal{K}_1\langle x,x'\rangle + \mathcal{K}_2)$

It should be noted that the main feature of support vector machine (SVM) focuses on non-model assumption for the optimization. For instance, it is comparable for SVM linear boundary versus SVM non-linear boundary, or SVM with different types of kernel functions. However, for data that have legitimate model assumptions, such as the normal model for Bayesian classifiers or binary model for logistic regressions, comparing them with SVM is like comparing apples with pears. In fact, the differences on fundamental assumptions of the methodology hinders the legitimacy of comparisons.

The use of duality theorem makes SVM method switches the optimization process from dealing with the feature space X to the output space on the response variable y. This partially makes SVM popular in high dimension data analysis, such as in picture recognition for artificial intelligence.

8.4 Duality Theorem with Perturbation

The critical device in the operation of support vector machine (SVM) is the duality theorem in the optimization process. When the feasible solution can not be explicitly obtained in linear programming, the technique can be further advanced by a perturbation method in linear programming. In what follows in this section, we shall discuss a perturbation method in seeking the optimal solution for a bivariate Bonferroni lower bound with information on degree-two Bonferroni summations S_{11}, S_{21}, S_{12}, and S_{22}, defined as

$$S_{r,u} = \begin{cases} \sum_{\mathcal{U}} P(A_{i_1}...A_{i_r} B_{j_1}...B_{j_u}) & 1 \leq r \leq n, 1 \leq u \leq m \\ S_u(B) & r = 0, \quad 0 \leq u \leq m \\ S_r(A) & u = 0, \quad 0 \leq r \leq n \\ 1 & r = u = 0 \\ 0 & r > n \quad or \quad u > m, \end{cases}$$

where $\mathcal{U} = \{1 \leq i_1 < ... < i_r \leq n \quad 1 \leq j_1 < ... < j_u \leq m\}$, for any two sets of events A_1, ..., A_n and B_1, ..., B_m in an arbitrary probability space $(\Omega, \mathcal{F}, P)$.

The use of duality theorem for optimization solution in prediction essentially stems from the same root of optimization in linear programming. We briefly introduce the roadmap of perturbation method in this section. More details on the use of the perturbation method with applications can be found in the book by Chen (2014) [22].

For any two sets of events $\{A_i, i = 1, ..., n\}$ and $\{B_j, j = 1, ..., m\}$, let v_1 and v_2 be the number of occurrences of the two event sets, respectively. Denote $p_{ij} = P(v_1 = i, v_2 = j)$. For any integers $1 \leq t \leq n$ and $1 \leq k \leq m$, consider a set of consistent bivariate Bonferroni summations S_{ij}, $i = 1, ..., t$, $j = 1, ..., k$.

Similar to the optimization process in the construction of support vector machine (SVM), an optimal upper bound for $P(v_1 \geq 1, v_2 \geq 1)$ is defined by the maximum value of the following linear programming problem:

$$\max(p_{11} + ... + p_{nm}) \tag{8.1}$$

subject to $\sum_{i=0}^{n} \sum_{j=0}^{m} p_{ij} = 1$, and

$$p_{11} + 2p_{21} + ... + tp_{t1} + ... + nmp_{nm} = S_{11}$$
$$p_{21} + ... + \binom{t}{2}p_{t1} + ... + \binom{n}{2}mp_{nm} = S_{21}$$
$$...... \quad ... \quad ...$$
$$p_{tk} + ... + \binom{n}{t}\binom{m}{k}p_{nm} = S_{tk}.$$

The following is the rationale of optimality behind the bounding formulation in (8.1). For any solution $\mathbf{p}^*$ to the optimization issue (8.1), denote the sum of partial elements of $\mathbf{p}^*$, we have

$$U^* = \sum_{i=1}^{n} \sum_{j=1}^{m} p_{ij}^*.$$

Now for any two sets of events characterized by $\mathbf{p}$ in any probability space,

$$P(v_1 \geq 1, v_2 \geq 1) = \sum_{i=1}^{n} \sum_{j=1}^{m} p_{ij} \leq U^*,$$

because of (8.1). Thus the feasible optimal solution in (8.1) leads to a probability upper bound on the probability of the joint event

$$P(v_1 \geq 1, v_2 \geq 1).$$

Denote $T = \{(i,j) : i =, ..., t, j = 1, ..., k\}$. Let F_1 be an upper bound such that for any particular set of consistent Bonferroni summations S_{ij}, $(i,j) \in T$. If there exists a set of events $A_1^*, \ldots, A_n^*, B_1^*, \ldots, B_m^*$ in a probability space where

$$S_{ij}(A_1^*, \ldots, A_n^*, B_1^*, \ldots, B_m^*) = S_{ij}, \quad (i,j) \in T$$

and

$$P(v_1(A^*) \geq 1, \quad v_2(B^*) \geq 1) = F_1(S_{ij}, (i,j) \in T), \tag{8.2}$$

then F_1 is said to be a Fréchet optimal upper bound for $P(v_1 \geq 1, \quad v_2 \geq 1)$. References on Fréchet optimal bounds can be found in [102], among others.

The optimality for probability bounds defined in (8.2) can be translated into the language of linear programming, in which the domain of the function class, F_1, is limited to S_{ij}, a linear combination of p_{ij}. Denote,

$$\mathbf{b} = (\mathbf{1}, \mathbf{t}(1,1), \mathbf{t}(1,2), \mathbf{t}(2,1), \mathbf{t}(2,2))'\mathbf{x} \tag{8.3}$$

where $\mathbf{1}$ is the vector with length 2^{n+m} and all elements equal 1, and $\mathbf{t}(i,j)$ is the vector specified below for $i, j = 1, 2$,

$$\mathbf{t}'(i,j) = (1, ij, \frac{ij(j-1)}{2}, \frac{ij(i-1)}{2}, \frac{ij(i-1)(j-1)}{4}). \tag{8.4}$$

Denote the matrix

$$\mathbf{R} = (\mathbf{1}, \mathbf{t}(1,1), \mathbf{t}(1,2), \mathbf{t}(2,1), \mathbf{t}(2,2))',$$

a $(5 \times 2^{n+m})$ matrix with structure not affected by the values of the S_{ij}'s. We have

$$\mathbf{b} = \mathbf{R}\mathbf{x}. \tag{8.5}$$

Vectorization of a bivariate array.   Letting $w = (n+1)(m+1)$ and putting w elements of p_{ij} for $i = 0, 1, ..., n$ and $j = 0, 1, ..., m$ into a vector of length w, yields

$$\mathbf{q} = (p_{ij})_{w \times 1}$$

where p_{ij} are arranged from $p_{0,0}$ to p_{nm} by the increasing order on i for each value of j, and then on j for each increasing sequence of i. Such a method of vectorization for a bivariate array can be illustrated by the following example. Letting $n = m = 2$, the vectorized outcome of a 9×1 vector $\mathbf{q}$ reads

$$\mathbf{q} = (p_{00}, p_{10}, p_{20}, p_{01}, p_{11}, p_{21}, p_{02}, p_{12}, p_{22})'.$$

Furthermore, by the expression of bivariate Bonferroni summations, see for example, Chen(2014)[22].

$$S_{kt} = \sum_{i=k}^{n} \sum_{j=t}^{m} \binom{i}{k} \binom{j}{t} p_{ij} = \mathbf{g}'_{kt}\mathbf{q}, \quad k = 1, 2 \quad t = 1, 2, \qquad (8.6)$$

where the row vector $\mathbf{g}'_{kt}$ is the vector of coefficients specified in (8.6). Combining the row vectors $\mathbf{g}'_{k,t}$ into a matrix $\mathbf{G}$ (with the first row as $\mathbf{1}$) for the quantities $S_{11}, S_{12}, S_{21}, S_{22}$.

Therefore putting $\mathbf{b}' = (1, S_{11}, S_{12}, S_{21}, S_{22})$, for $w = (n+1)(m+1)$, there exists a $5 \times w$ matrix $\mathbf{G}$ so that

$$\mathbf{b} = \mathbf{Gq}, \qquad (8.7)$$

where the first row of the matrix $\mathbf{G}$ is $\mathbf{1}'$, the first $(m+1)th$ column of $\mathbf{G}$ is $(1, 0, 0, 0, 0)'$, and the structure of $\mathbf{G}$ is not affected by the value of the bivariate Bonferroni summations $S_{i,j}$'s.

Denote the vector $\mathbf{c}$: $\mathbf{c} = (c(i, j))$, with the first element of $\mathbf{c}$ corresponding to the index $\{i = 0 \text{ or } j = 0\}$, and the rest of the elements of $\mathbf{c}$ formed by ranking over $i \geq 1$ in increasing order for each fixed j, then over $j \geq 1$ in increasing order. Also assign

$$c(i, j) = \begin{cases} 0, & for \ i{=}0 \ or \ j{=} 0 \\ 1, & otherwise. \end{cases}$$

The joint probability of at least one occurrence in both event sets can be expressed as

$$P(v_1 \geq 1, \ v_2 \geq 1) = \mathbf{c}'\mathbf{p}.$$

With the setting above, we have the following theorems.

Theorem 8.1 *For matrix* $\mathbf{G}$, *vectors* $\mathbf{c}$ *and* $\mathbf{b}$ *as specified in the above setting, denote the vector* $\mathbf{w}' = (w_0, w_1, w_2, w_3, w_4)$ *which may depend on* m, n, *but not on the values of* S_{11}, S_{12}, S_{21}, S_{22} *under consideration.* $\mathbf{w}'\mathbf{b}$ *is an upper bound for* $P(v_1 \geq 1, \ v_2 \geq 1)$ *for all probability spaces if and only if* $\mathbf{w}'\mathbf{G} \geq \mathbf{c}'$ *(each element of the vector* $\mathbf{w}'\mathbf{G}$ *is not less than the corresponding element in the vector* $\mathbf{c}$*).*

The following theorem explores the existence of the feasible optimal solution from the angle of probability theory, instead of linear programming.

Theorem 8.2 *If the Bonferroni summations S_{11}, S_{12}, S_{21}, and S_{22} are consistent, the linear programming upper bound for $P(v_1 \geq 1, v_2 \geq 1)$ always exists.*

Similar to the argument in the preceding section, to optimize the target function, we find a vector (probability space) $\mathbf{p}$ to minimize $\sum_{i=1}^{n} p_i$ subject to $\mathbf{p} \geq \mathbf{0}$ and $\mathbf{Gp} = \mathbf{b}$, where

$$\mathbf{b} = (1, S_{11}, S_{12}, S_{21}, S_{22})'.$$

Here, $\mathbf{G}$ is a $5 \times (nm + 1)$ matrix with a typical column $\mathbf{a}_t$, where the general form of $\mathbf{a}_t$ is:

$$\mathbf{a}_t' = (1, ij, \frac{ij(j-1)}{2}, \frac{ij(i-1)}{2}, \frac{ij(i-1)(j-1)}{4}), \qquad (8.8)$$

for integers $1 \leq i \leq n$ and $1 \leq j \leq m$.

For matrix $\mathbf{G}$ defined above, for a vector of consistent Bonferroni summations $\mathbf{b}$, and a vector of coefficients $\mathbf{c}$, denote the vector

$$\mathbf{w}' = (w_0, w_1, w_2, w_3, w_4)$$

which may depend on m, n, but not on the values of S_{11}, S_{12}, S_{21}, S_{22}.

Theorem 8.3 *The value $\mathbf{w}'\mathbf{b}$ is a lower bound for $P(v_1 \geq 1, v_2 \geq 1)$ for all probability spaces if and only if $\mathbf{w}'\mathbf{A} \leq \mathbf{c}'$ (each element of the vector $\mathbf{w}'\mathbf{A}$ is not greater than the corresponding element in the coefficient vector $\mathbf{c}$).*

The above theorem leads to the existence of degree-two optimal probability lower bound for the occurrence of at least one joint event, as stated below. More details on the proofs of the theorem can be found in Chen(2014)[22].

Theorem 8.4 *For a set of consistent Bonferroni summations S_{11}, S_{12}, S_{21}, and S_{22}, the linear programming lower bound for $P(v_1 \geq 1, v_2 \geq 1)$ always exists.*

Now that the existence is proved above, the following describes a perturbation method to find the bivariate optimal lower bound. If the condition in Theorem 8.3 is satisfied, the optimal lower bound is found. However, it is not always true that $\mathbf{x_B} = \mathbf{B}^{-1}\mathbf{b} \geq \mathbf{0}$. When $\mathbf{x_B} = \mathbf{B}^{-1}\mathbf{b} \ngeq \mathbf{0}$, we need to find an alternative approach to reach the linear programming bound. This is technically more involved in linear programming. In the following, we provide details on the alternative approach (an iteration algorithm) and show that the algorithm can theoretically reach the existence condition after finite iterations.

When $\mathbf{B}^{-1}\mathbf{b} \not\geq \mathbf{0}$, Theorem8.3 cannot be applied to find the linear programming lower bound. We can update the matrix B by replacing one of the columns in $\mathbf{B}$ with a column from the set of columns of $\mathbf{G}$ to form a new $\mathbf{B}$ matrix, and denote the updated $\mathbf{B}$-matrix as $\mathbf{B}_1$. If $\mathbf{B}_1^{-1}\mathbf{b} \geq \mathbf{0}$, by Theorem 8.3, we find the linear programming bound. If $\mathbf{B}_1^{-1}\mathbf{b} \not\geq \mathbf{0}$, a column in $\mathbf{B}_1$ is replaced to form a new $\mathbf{B}$-matrix, denoted as $\mathbf{B}_2$. In this way, we initiate an iteration process.

The hug of the device now is in selecting the proper column from $\mathbf{G}$ to form the optimal point. To this end, we use a perturbation device to show that once the column for removal/replacement is selected appropriately, we can achieve the solution for a linear programming lower bound.

The introduction of the ϵ below, ostensibly a perturbation device, is necessary to show that cycling doesn't occur in the iterative procedure aimed at arriving at an optimal lower bound by a linear programming implementation.

For any positive value $\epsilon > 0$, define a vector $\mathbf{c}_B(\epsilon)$ as follows. Let $\Gamma = \{\mathbf{c}(\epsilon)\}$, where $\mathbf{c}(\epsilon) = (c_1(\epsilon), c_2(\epsilon), ..., c_{nm+1}(\epsilon))'_{(nm+1)\times 1}$ is any vector of ϵ's satisfying,

$$
\begin{aligned}
c_1(\epsilon) &= 1 \\
c_i(\epsilon) &= \epsilon^{k(i)} \quad \text{for some} \quad k > 0 \quad \text{depending on} \quad i > 1 \qquad (8.9)\\
c_i(\epsilon) &\neq c_j(\epsilon) \quad i \neq j.
\end{aligned}
$$

For example, the function of ϵ can take any one of the following forms,

$$c_i(\epsilon) = \epsilon^i, i \neq 1, \quad c_1(\epsilon) = 1$$

or

$$c_i(\epsilon) = \epsilon^{i+7}, i \neq 1, \quad c_1(\epsilon) = 1.$$

The function $\mathbf{c}(\epsilon)$ will be fixed through the iteration process. The possibility of different explicit forms of $\mathbf{c}(\epsilon)$ is key in showing that it is not necessary to worry about perturbation terms in ϵ.

Now, we can use columns from $\mathbf{G}$ to form $\mathbf{B}$ denoted as $\mathbf{a}_1, \mathbf{a}_{t_2}, ..., \mathbf{a}_{t_5}$, then $\mathbf{c}_B(\epsilon) = (1, c_{t_2}(\epsilon), ..., c_{t_5}(\epsilon))'$, a 5×1 vector formed by selecting the corresponding $c_i(\epsilon)$'s from the $(nm+1) \times 1$ vector $\mathbf{c}(\epsilon)$ introduced above.

With the vector $\mathbf{c}_B(\epsilon)$ defined above, we have

$$\mathbf{c}_B(\epsilon)'\mathbf{B}^{-1}\mathbf{a}_k \geq c_k(\epsilon), \quad k \neq 1; \quad \text{and} \quad \mathbf{c}_B(\epsilon)'\mathbf{B}^{-1}\mathbf{a}_1 = 1, \qquad (8.10)$$

where $\mathbf{a_1}, ..., \mathbf{a_k}$ are columns of matrix $\mathbf{G}$.

Assume matrix $\mathbf{B}$ defined above satisfies the conditions specified in (8.10) for any $0 < \epsilon \leq \epsilon_0$. If the vector $\mathbf{x_B} = \mathbf{B}^{-1}\mathbf{b} \geq \mathbf{0}$, then there exists a lower bound of $P(v_1 \geq 1, v_2 \geq 1)$ for any probability space $\mathbf{p}$, and there exists a probability space in which this lower bound achieves equality.

Denote

$$
\mathbf{B_1} = \begin{pmatrix}
1 & 1 & 1 & 1 & 1 \\
0 & ab & (a+1)b & a(b+1) & (a+1)(b+1) \\
0 & \frac{ab(b-1)}{2} & \frac{(a+1)b(b-1)}{2} & \frac{ab(b+1)}{2} & \frac{(a+1)(b+1)b}{2} \\
0 & \frac{ab(a-1)}{2} & \frac{(a+1)ab}{2} & \frac{a(a-1)(b+1)}{2} & \frac{(a+1)(b+1)a}{2} \\
0 & \frac{ab(b-1)(a-1)}{4} & \frac{(a+1)ab(b-1)}{4} & \frac{ab(a-1)(b+1)}{4} & \frac{ab(a+1)(b+1)b}{4}
\end{pmatrix}
$$

then, as shown in Chen (2014)[22], there exists a positive integer ϵ_1 so that the matrix $\mathbf{B_1}$ satisfies condition (8.10) for all $0 < \epsilon < \epsilon_1$.

Notice that the inverse matrix of $\mathbf{D}$ can be decomposed as follows.

$$
\mathbf{D}^{-1} = \mathbf{B}^{-1} - \begin{pmatrix} \frac{y_{1j_k}}{y_{rj_k}}\mathbf{s_r}' \\ \vdots \\ \frac{y_{5j_k}}{y_{rj_k}}\mathbf{s_r}' \end{pmatrix} + \begin{pmatrix} \mathbf{0}' \\ \vdots \\ \frac{1}{y_{rj_k}}\mathbf{s_r}' \\ \vdots \\ \mathbf{0}' \end{pmatrix}.
$$

Let n be the total number of iterations before reaching the optimal solution, if $\theta(\epsilon) < 0$ and the condition (8.10) persists for each $\mathbf{B}$ matrix in the iteration process, $n < \infty$.

Proof: Put $z_{rb} = \mathbf{s_r}' \mathbf{b}$, we have

$$
\mathbf{D}^{-1}\mathbf{b} = \mathbf{B}^{-1}\mathbf{b} - \begin{pmatrix} \frac{y_{1j_k}}{y_{rj_k}}z_{rb} \\ \vdots \\ \frac{y_{5j_k}}{y_{rj_k}}z_{rb} \end{pmatrix} + \begin{pmatrix} 0 \\ \vdots \\ \frac{z_{rb}}{y_{rj_k}} \\ \vdots \\ 0 \end{pmatrix}.
$$

Thus,

$$
\begin{aligned}
\mathbf{c_D}'(\epsilon)\mathbf{D}^{-1}\mathbf{b} &= \mathbf{c_B}'(\epsilon)\mathbf{B}^{-1}\mathbf{b} - \theta(\epsilon)\mathbf{s_r}'\mathbf{b} \\
&< \mathbf{c_B}'(\epsilon)\mathbf{B}^{-1}\mathbf{b},
\end{aligned}
$$

since $\theta(\epsilon) < 0$ and $\mathbf{s_r}'\mathbf{b} < 0$ for $x_r < 0$.

Thus $f(\mathbf{B}) = \mathbf{c_B}(\epsilon)'\mathbf{B}^{-1}\mathbf{b}$ decreases strictly at the next iteration, for $0 < \epsilon < \epsilon_0(\mathbf{B})$. The number of permissible $\mathbf{B}$'s is however, finite, (at most $nm(nm-1)(nm-2)(nm-3)$), so there will be a stage in the iteration process where no strict decrease is possible. But strict decrease is possible at the next stage for any $\mathbf{B}$ satisfying (8.10) and $\mathbf{B}^{-1}\mathbf{b} \not\geq \mathbf{0}$. Thus at some iteration $\mathbf{B}^{-1}\mathbf{b} \geq \mathbf{0}$.

This means that after a finite number of iterations (with the first column of $\mathbf{B}$ never being changed), there exists a matrix $\mathbf{B}$ and an associated $\theta(\epsilon)$ such that for all $0 < \epsilon < \epsilon_0(\mathbf{B})$.

Summarizing the above discussion, we have the following conclusions.

- For any $\mathbf{B}$ satisfying (8.10), and $\mathbf{x}_B = \mathbf{B}^{-1}\mathbf{b} \not\geq \mathbf{0}$, let r represent the position of the smallest negative element of $\mathbf{x_B} = \mathbf{B}^{-1}\mathbf{b}$, then $r > 1$. Furthermore, there exists a set of vectors $H(\mathbf{B}) = \{\mathbf{a}_{j_1}, ..., \mathbf{a}_{j_t}\}$ for some $t \geq 1$, in $\mathbf{A}$, such that $\mathbf{s}_r{}'\mathbf{a}_{j_1} < 0$, ..., $\mathbf{s}_r{}'\mathbf{a}_{j_t} < 0$, and $\mathbf{b}_i \notin H(\mathbf{B})$ where $\mathbf{b}_i$ is the ith column of $\mathbf{B}$.

- For a matrix $\mathbf{B}$ satisfying (8.10), if $\mathbf{x}_B = \mathbf{B}^{-1}\mathbf{b} \not\geq \mathbf{0}$, the corresponding $\theta(\epsilon) < 0$, for all $0 < \epsilon < \epsilon_2$, with a value $\epsilon_2 = \epsilon_2(\mathbf{B}) > 0$. Also, the vector $\mathbf{a}_{j_k}$, which is the maximizing vector $\mathbf{a}_p$, $\quad p \in \{j_1, ..., j_t\}$, associated with $\theta(\epsilon)$ for $0 < \epsilon < \epsilon_2(\mathbf{B})$, is uniquely determined.

Chen (2014)[22] also shows the following results to ensure the smoothing operation in the iterative process.

Theorem 8.5 *The condition $\theta(\epsilon) < 0$ in each iteration, which means that degeneracy does not occur in the iteration process. In the iteration process, when we sequentially reach a matrix $\mathbf{B}$ such that $\mathbf{B}^{-1}\mathbf{b} \geq \mathbf{0}$, an optimal solution is found and the process is stopped. If the associated $\mathbf{B}^{-1}\mathbf{b} \not\geq \mathbf{0}$, Condition (8.10) persists for every $\mathbf{B}$ used in each iteration, with a corresponding $\epsilon_0(\mathbf{B}) > 0$.*

More details on the theory of perturbation method can be found in Chapter 4 of [22].

Summary: This chapter discusses the method of support vector machine, a classification methodology that does not rely on model assumptions for the distribution of the data. We start with the simplest case of *maximal margin classifier*, where the data can be linearly separated into two distinct categories. In the scenario where some observations can not be conveniently classified, we discuss the method of support vector classifier, linear classifier with permitted misclassified support vectors. When the kernel function of the classification is extended to non-linear function with misclassified support vectors, the method of support vector machine (SVM) is discussed.

The methodology of SVM is grounded upon the duality theorem in the process of seeking the optimal solution for a linear programming problem [39]. In the case where explicit feasible solution is not available, we illustrate a perturbation method to screen for the optimal solution. The overall idea works similarly to the greedy search algorithm. We add a perturbation term in each step for a local optimization, then find the shortest way to reach the largest increase of the distance measure repeatedly, and prove that the iteration will converge to the global optimal solution.

References for the optimal lower probability bound using the Bonferroni summations can be found in [17], [22], [34], [50] [62], [65], [71], [100], among others. These bounds can be improved with duality theorem as discussed in this chapter. When high-dimensional probability bounds are of interest, the bivariate inequality can be extended to a multivariate version in linear programming. Related references include [8], [22], [32], [31], [33], [71], [104], [106], among others. Without the condition of linear combination on Bonferroni summations, the perturbation method can be expanded to improve non-linear bounds such as inequalities documented in [22], [19], [63] [64], [87], and [89], [101], among others. The derivation for optimal solutions on SVM is similar to the derivation of the linear programming bounds in probability inequalities. More applications in this regard can be found in [22], [61], [90] [103], [107], and [108], among others.

9

Decision Trees and Range Regressions

Consider the prediction of an outcome (numerical or categorical variable) on the basis of a predictor vector featured by p variables, $\mathbf{x} = (X_1, ..., X_p)'$. Instead of assuming a point-wise relationship between y and $\mathbf{x}$ such as in linear regression

$$y = \alpha + \beta' \mathbf{x} + \epsilon,$$

the method of decision trees or range regression seeks the connection between a proxy of Y and a partition of the feature space $\mathbf{x} \in R^p$. We start with an illustrative example.

Example 9.1 Age impact on systolic blood pressure. *Assume that we are interested in predicting the systolic blood pressure using patient information on age and gender. Denote X_1, X_2, and y as age, gender, and systolic blood pressure, respectively, with data information in Table 9.1 below.*

TABLE 9.1

Systolic blood pressure dataset

Age	Gender	Systolic BP
20	M	112
17	M	102
19	F	138
15	F	142
40	M	164
53	M	158
51	F	153
42	F	167

Using linear regression, the fitted model based on data information in Table 9.1, reads,

$$Systolic\ BP = 106.5 + 1.106 * age + \varepsilon$$

The above linear regression line is easy to interpret. It implies that, on average, the SBP increases 1.106 units per year of age increase. This reflects the possible effect of bulging veins (when people get older, the valves in the veins may wear out and result in improper blood flow in the extremities back to the heart). However, it is not necessary that the change of SBP is based on per unit increase of age. According to the model, an eighty-year-old patient might get, on average, a SBP at the level of

$$106.47 + 1.106 * 80 = 194.95,$$

which is obviously unrealistic. In fact, it would be more plausible to assume that for a certain age range, the SBP changes to a common value. For instance, the CDC website states the following relationship.

age range	systolic blood pressure
14-18	90-120
19-40	95-135
41-60	110-145
61-older	95-145

The above idea can be formulated into the following equation:

$$y = f(X) = \sum_{i=1}^{k} c_i I(X, R_i),$$

where R_1, ..., R_k are age intervals of the patient, and

$$I(X, R_i) = \begin{cases} 1 & if \ X \in R_i \\ 0 & if \ X \notin R_i. \end{cases}$$

Definition 9.1 *Consider a prediction problem with a response variable Y and predictor $\mathbf{x} = (X_1, ..., X_k)' \in S$, where S is the feature space. A binary splitting on element X_j, $j \in \{1, ..., k\}$, is a numerical value s that splits the feature space S into two sets*

$$R_1(j, s) = \{\mathbf{x} | X_j < s\} \qquad and \qquad R_2(j, s) = \{\mathbf{x} | X_j \geq s\},$$

so that

$$R_1(j, s) \bigcup R_2(j, s) = S.$$

The concept of binary splitting forms a partition of the feature space S. With that, we can define a decision tree as follows.

Definition 9.2 *Consider a prediction problem with a response variable Y and predictor $\mathbf{x} \in S$. Let R_1, .., R_k be a partition of the feature space generated by recursively splitting the domain of elements of $\mathbf{x}$, with optimization of the homogeneity measure at each step of the binary split. A decision tree is a relationship that divides the feature space into groups according to a homogeneity measure,*

$$\hat{f}(X) = \sum_{i=1}^{k} c_i I(X, R_i),$$

where c_i is the common outcome for $\mathbf{x}$ with features in the partition R_i.

We introduce the concept *homogeneity measure* in the above definition. When the response variable is continuous, the usual mean squared error

$$\frac{1}{n} \sum_{i=1}^{n} (y_i - \hat{y}_i)^2$$

is the homogeneity measure for the *closeness* between the observed and the predicted outcomes. However, when the response is categorical, the value of $(white - green)^2$ has no meaning for comparisons. Even when we label the categories with numbers such as *1 for red, 2 for green* and *3 for white*, the outcome

$$(1 - 2)^2 < (1 - 3)^2$$

does not represent the color differences among red, green, and white. In this case, we need to use other homogeneity measure such as Gini index or entropy described in Section 9.2.

The selection of homogeneity measure is dictated by the type of response variable Y. In what follows in this chapter, we shall redirect our discussion to regression trees and classification trees, separately. The chapter will be concluded by a discussion on range regression, an extension of binary splitting in regression trees into multiple splitting.

9.1 Regression Trees and UMVUE

When the response variable is continuous, the distance is a usual measure on homogeneity. Features with similar responses can be measured by the distance on RSS (residual sum of squared errors).

$$RSS = \sum_{i:x_i \in R_1(j,s)} (y_i - \hat{y}_{R_1})^2 + \sum_{i:x_i \in R_2(j,s)} (y_i - \hat{y}_{R_2})^2,$$

where $R_1(j,s) = \{x | X_j < s\}$ and $R_2(j,s) = \{x | X_j \geq s\}$.

Notice that a regression tree is formed via recursive binary splitting of the feature space with RSS as the homogeneity measure. We use the following hypothetical example to illustrate the process in the construction of a regression tree.

Example 9.2 *Consider the following simple data set for the construction of a regression tree.*

x	1.5	2.6	5.1	9.2
y	2	1	10	20

In the first step, possible splitting points for the binary splitting of x are

$$\frac{1.5 + 2.6}{2} = 2.05, \qquad \frac{2.6 + 5.1}{2} = 3.85, \qquad \frac{5.1 + 9.2}{2} = 7.15,$$

the corresponding RSS reads

splitting point	2.05	3.85	7.15
RSS	181.29	50.5	48.67

Since the splitting point at 7.15 has the smallest RSS, the first binary splitting is

$$R_1 = \{x | x < 7.15\}, \qquad R_2 = \{x | x \geq 7.15\}.$$

After the first binary splitting, since R_2 contains only one observation, the second binary splitting occurs at a value $s2$ that minimizes the RSS among the rest of the three observations. Similar to the first split, we have

s2	2.05	3.85
RSS	40.5	0.5

Thus, the second binary split is $s2 = 3.85$.

Using R packages, the constructed tree can be seen in Figure 9.1.

It should be noted that for each partition on the feature space, when the distribution family of the sample mean of y is a complete distribution family, the UMVUE (uniformly minimum variance unbiased estimator) of the mean response at each terminal node of the regression tree is

$$\hat{c}_i = \bar{y}_i = \frac{1}{|R_i|} \sum_{\mathbf{x}_j \in R_i} y_j,$$

which is the average of all responses that share the corresponding features $\mathbf{x} \in R_i$.

By the consistency of UMVUE and the law of large numbers, when the number of observations in each terminal node is large enough, the sample mean can always approximate the population mean. However, in practice, we do not always have a large sample size in each terminal node in regression trees. It should be noted that with limited sample sizes, the sample mean is not always the UMVUE of the mean parameter, as shown in the following example.

Example 9.3 *Consider a random sample X_1, ..., X_n from a Uniform model $U(0, \theta)$. $E(X) = \frac{\theta}{2}$, so the sample mean $\bar{X}$ is an unbiased estimate of $\frac{\theta}{2}$. However, $\bar{X}$ is not complete in this case, so it is not UMVUE of the population mean.*

On the other hand, the estimator $T = X(n)$, the largest order statistic, is complete and

$$E(T) = \frac{n}{n+1} \frac{\theta}{2},$$

```
x<-c(1.5, 2.6, 5.1, 9.2)
y<-c(2, 1, 10, 20)
library(rpart)
tt<-data.frame(x,y)
rt <- rpart(y~x, minsplit=2, data=tt, method="anova")
par(xpd=NA)
plot(rt)
text(rt, use.n="FALSE", all=FALSE)
```

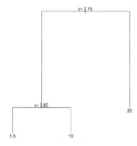

FIGURE 9.1

Hypothetical data for regression tree construction

thus the UMVUE of the population mean $\theta/2$ is

$$\hat{\mu} = \frac{\hat{\theta}}{2} = \frac{n+1}{n}X(n),$$

which is not the sample mean $\bar{X}$.

Thus, in the construction of regression trees for continuous responses, on the basis of the characteristics of the response, properly adjusting the estimate for the constant in the terminal node will increase the accuracy and convergence rate for the regression tree.

Another key step in the construction of a regression tree accounts for the time to stop the binary splitting of the feature space. The continuation of binary splitting will eventually break the data points into individual cells where each terminal node contains only one observation. This is not desirable because it over-fits the relationship between the response Y and the feature $\mathbf{x}$. To this end, the construction of a regression tree necessitates a cost-complexity pruning step. Notice that for each value of α and a tree T_0, there exists a subtree $T \subset T_0$, such that

$$R = \sum_{m=1}^{|T|} \sum_{x_i \in R_m} (y_i - \hat{y}_{R_m})^2 + \alpha |T|$$

is as small as possible, where R_m is the m-th terminal node. Hence, T is the selected tree for the data. Here, the value of α can be determined using K-fold cross-validation.

1 Select a set of α values as candidates. For each α value do the following.

2 For any positive integer $j \le K$, use recursive binary splitting to grow a tree on the cross-validation data, which is the training data except the j-th fold.

3 Use the cost complexity pruning to find a sub-tree T so that the term,

$$R = \sum_{m=1}^{|T|} \sum_{x_i \in R_m} (y_i - \hat{y}_{R_m})^2 + \alpha |T|,$$

is the smallest possible value, where R_m is the mth terminal node.

4 Compute the mean squared prediction errors on the data in the left-out jth fold.

5 Average the mean squared prediction error after running all j-fold data in the cross-validation process.

6 After computing the average cross-validation mean squared prediction error for all candidate α values, pick the α level that is associated with the smallest average of the mean squared prediction error.

9.2   Classification Tree

For the dataset $(\mathbf{x}, y)$ where y is a categorical variable representing different classes such as Red, Black, Green for colors, or Hypertension, Diabetes, Stroke, Cancer for diseases, the method of regression trees can not be directly applied. This is because the Euclidean distance (mean squared prediction error) is no longer a valid measure for category homogeneity. For instance, when we denote $Red = 1$, $Green = 2$, and $Black = 3$, the error misclassifying Red as $Green$ (which reads MSE $(1 - 2)^2 = 1$) is not less severe than the error misclassifying Red as $Black$ (which reads MSE $(1 - 3)^2 = 4$). In the literature, homogeneity measures for classification trees include misclassification rate, Gini index, and entropy. Toward this end, the decision tree definition for classification problems can be adjusted as follows.

Definition 9.3 *Consider a prediction problem with response variable $Y \in \{1, ..., K\}$ and predictor $\mathbf{x} \in S$. Let $R_1, .., R_m$ be a partition of the feature space generated by recursively splitting the domain of elements of $\mathbf{x}$ with optimization of the homogeneity measure at each step of the binary split. A classification tree is a relationship that divides the feature space into groups according to a homogeneity measure (misclassification rate, gini index, or entropy),*

$$\hat{f}(X) = \sum_{i=1}^{m} c_i I(X, R_i),$$

where $c_i \in \{1, ..., k\}$ is the proxy for the common value of observations $\mathbf{x}$ with features in the partition R_i.

9.2.1 Misclassification Rate

For classification problems, let $\hat{p}_{jk}$ denote the proportion of training observations in the jth region that belong to the kth class, $k \in \{1, 2, ..., K\}$ in each binary splitting, $j = 1, 2$. The misclassification rate for region j is

$$E_j = 1 - \max_{k \in \{1,...,K\}} (\hat{p}_{jk}).$$

Example 9.4 *When SBP (systolic blood pressure) is used to diagnose DVT (deep vein thrombosis), assume that features of 20 patients under the study can be described according to the following table.*

SBP vs DVT	DVT (Yes)	DVT (No)
SBP < 120	5	7
SBP ≥ 120	6	2

Out of the total of 20 patients, when 120 is selected as the threshold in the binary splitting of SBP, we have patients with $SBP < 120$ as the group $j = 1$, and patients with $SBP \geq 120$ as the group $j = 2$. Out of the 12 patients with $SBP < 120$, there are 5 with DVT ($k = 1$) and 7 without DVT ($k = 2$). Thus we have, for $j = 1$,

$$\hat{p}_{11} = \frac{5}{12} \qquad \hat{p}_{12} = \frac{7}{12},$$

and

$$E_1 = 1 - \max(\hat{p}_{11}, \hat{p}_{12}) = 1 - \max(\frac{5}{12}, \frac{7}{12}) = \frac{5}{12}.$$

So, Region-1 is labeled as no-DVT with misclassification rate $\frac{5}{12}$.

As for Region-2, $j = 2$, we have

$$\hat{p}_{21} = \frac{6}{8} \qquad \hat{p}_{22} = \frac{2}{8}$$

and

$$E_2 = 1 - \max(\frac{6}{8}, \frac{2}{8}) = \frac{2}{8}.$$

So, Region-2 is for patients with DVT and the misclassification rate is $\frac{2}{8}$.

Therefore, the overall classification rate for the binary split at $SBP = 120$ reads

$$E = \frac{5}{12} * \frac{12}{20} + \frac{2}{8} * \frac{8}{20} = \frac{7}{20}.$$

In fact, when we use the majority vote at each terminal node to determine the common class for the node, the misclassification rate is simply the sum of the misclassified proportions in the two regions of the binary splitting.

9.2.2   Gini Index

Instead of directly using the misclassification rate as a measure for the homogeneity in each terminal node, another measure is the Gini index, which measures the variation (or node purity) in the binary splitting during the construction of a classification tree. For each binary splitting, denote $\hat{p}_{jk}$ the proportion of training observations in the jth region that are from the kth class, $k \in \{1, 2, ..., K\}$, the Gini index reads,

$$\hat{G} = \sum_{k=1}^{K} \hat{p}_{jk}(1 - \hat{p}_{jk}).$$

For the n_j observations in the terminal node, denote $X_{jk} = 1$ with probability p_{jk} in the jth region for the observations that belong to Category k. We have

$$Var(X_{ijk}) = p_{jk}(1 - p_{jk}) \quad i = 1, ..., n_j,$$

and the population-wise Gini index for region j reads,

$$G(j) = \sum_{k=1}^{K} p_{jk}(1 - p_{jk}) = \sum_{k=1}^{K} Var(X_{jk}).$$

Thus, for each splitting in the region j, the Gini index $G(j)$ is essentially the sum of variance of each category in the region. In general, the Gini index for a value in a binary splitting reads

$$Gini(s) = \frac{n_1}{n}G(1) + \frac{n_2}{n}G(2),$$

where n_1 denotes the number of observations satisfying $X_j < s$, and n_2 denotes the number of observations satisfying $X_j \geq s$.

The Gini index takes on a small value if all of the correct classification errors are close to zero or one. It is a measure of node purity. A small value indicates that a node contains predominantly observations from a single class.

Example 9.5 *Using the SBP-DVT classification in Example 9.4, the Gini index in splitting SBP = 120 reads*
When j = 1,

$$Gini(1) = \frac{5}{12} * (1 - \frac{5}{12}) + \frac{7}{12} * (1 - \frac{7}{12}) = 2 * \frac{5}{12} * \frac{7}{12};$$

when j = 2,

$$Gini(2) = \frac{6}{8} * (1 - \frac{6}{8}) + \frac{2}{8} * (1 - \frac{2}{8}) = 2 * \frac{6}{8} * \frac{2}{8},$$

and the Gini index at SBP=120 reads

$$Gini(120) = 2 * \frac{5}{12} * \frac{7}{12} * \frac{12}{20} + 2 * \frac{6}{8} * \frac{2}{8} * \frac{8}{20} = \frac{53}{120}.$$

9.2.3 Entropy

For the previous measures of homogeneity in classification, the misclassification rate uses the percentage of node homogeneity. A small value of the misclassification rate indicates the predominant observations from a single class. The Gini index, on the other hand, focuses on the node purity via the variance of each class in the region. A small value of Gini index implies more observations in the node are from a single class. Besides these two measures of homogeneity, another commonly applied homogeneity measure for binary response is the measure of *entropy*.

Entropy is a concept used in information theory, where a higher probability of the occurrence of an event implies less information obtained when the event occurs; and a lower probability implies more information obtained when the event occurs. It is a useful concept in coding and decoding a signal process. The form $-plog(p)$ stems from the fact that the function

$$f(p) = -c * log_a(p)$$

for constants a and c is the only function satisfying the following three conditions in information theory.

1 $f(x)$ is a monotonically decreasing function of x;

2 $f(x)$ is continuous in x;

3 $f(p_1 \times p_2) = f(p_1) + f(p_2)$, which implies that information of two independent events is the sum of the individual information of each event.

On the basis of the above definition, when $I(X) = -log(P(X = x))$ is used for the self-information of the element x, the entropy of a discrete random variable X with probability mass function p_x is defined as the expected value of the information belonging to each basic element.

$$Entropy(X) = E(I(X)) = \sum_x (-log(p))P(X = x).$$

For each binary splitting, denote $\hat{p}_{jk}$ the proportion of training observations in the jth region that are from the kth class, $k \in \{1, 2, ..., K\}$, we have the entropy,

$$\hat{D} = -\sum_{k=1}^{K} \hat{p}_{jk} log[\hat{p}_{jk}].$$

Since the classification error is between 0 and 1, the entropy in a binary split is non-negative. It takes a value near zero if the classification rates are all near zero or near one.

It should be noted that the three different measures of node homogeneity assume different meanings when splitting in the construction of the classification tree, which should be integrated into the interpretation.

We use the following example to comprehensively explain the three measures of node homogeneity in the construction of classification trees.

Example 9.6 *Consider a sample of 800 objects featured by variable x to be classified into two categories A and B.*

Number	300	100	200	100	100
Category	A	A	B	B	B
x	9.1	10.9	29.1	10.9	9.1

On the basis of the values of x, the candidate splitting points are $x = 10$ and $x = 20$.

Case-1: Using misclassification rate as the measure of node homogeneity
At $x = 10$, we have

- when $x < 10$, the corresponding outcomes include A -300 and B-100, the majority vote classifies the node as Category-A.

- when $x \geq 10$, the corresponding outcomes include A -100 and B-300, the majority vote classifies the node as Category-B.

The overall misclassification rate for splitting over $x = 10$ reads

[100(misclassification of B as A)+100(misclassification of A as B)]/800 = 0.25.

Alternatively, for $x < 10$, $j = 1$, $k \in \{A, B\}$, $p_{1A} = 3/4$ and $p_{1B} = 1/4$,

$$E_1 = 1 - max(3/4, 1/4) = 1/4.$$

for $x \geq 10$, $j = 2$, $k \in \{A, B\}$, $p_{2A} = 1/4$ and $p_{2B} = 3/4$,

$$E_2 = 1 - max(3/4, 1/4) = 1/4.$$

and the overall misclassification rate reads,

$$\frac{1}{4}\frac{400}{800} + \frac{1}{4}\frac{400}{800} = 1/4.$$

At $x = 20$, we have

- when $x < 20$, the corresponding outcomes include A -400 and B-200, the majority vote classifies the node as Category-A.

- when $x \geq 20$, the corresponding outcomes include A - none and B-200, the majority vote classifies the node as Category-B.

The overall misclassification rate for splitting over $x = 20$ reads

$(200(\text{misclassification of B as A}) + 0(\text{misclassification of A as B}))/800 = 0.25.$

Alternatively, for $x < 20$, $j = 1$, $k \in \{A, B\}$, $p_{1A} = 4/6$ and $p_{1B} = 2/6$,

$$E_1 = 1 - max(2/3, 1/3) = 1/3.$$

for $x \geq 20$, $j = 2$, $k \in \{A, B\}$, $p_{2A} = 0$ and $p_{2B} = 200/200 = 1$,

$$E_2 = 1 - max(0, 1) = 1.$$

and the overall misclassification rate becomes

$$\frac{3}{4}\frac{1}{3} + \frac{1}{4} * 0 = 1/4.$$

There is a tie in the homogeneity measure with misclassification errors at 10 and 20. In this case, the tie can be broken by flipping a fair coin.

Case-2: Using Gini index as the measure of node homogeneity
At splitting point $x = 10$, we have

- when $x < 10$, the corresponding outcomes include A -300 and B-100, the majority vote classifies the node as Category-A. $\hat{p}_{1A} = 300/400$.

- when $x \geq 10$, the corresponding outcomes include A -100 and B-300, the majority vote classifies the node as Category-B. $\hat{p}_{2B} = 300/400$.

For $x < 10$, $j = 1$ the Gini index

$$G(1) = \frac{3}{4} * (1 - \frac{3}{4}) + \frac{3}{4} * (1 - \frac{3}{4}) = \frac{3}{8}.$$

For $x \geq 10$, $j = 2$ the Gini index

$$G(2) = \frac{1}{4} * (1 - \frac{1}{4}) + \frac{3}{4} * (1 - \frac{3}{4}) = \frac{3}{8}.$$

The overall Gini index at splitting $x = 10$ reads

$$Overall\ Gini = \frac{3}{8} * \frac{400}{800} + \frac{3}{8} * \frac{400}{800} = 3/8.$$

At splitting point $x = 20$, we have

- when $x < 20$, the corresponding outcomes include A - 400 and B -200, the majority vote classifies the node as Category-A. $\hat{p}_{1A} = 400/600$.

- when $x \geq 20$, the corresponding outcomes include A - none and B -200, the majority vote classifies the node as Category-B. $\hat{p}_{2B} = 200/200 = 1$.

For $x < 20$, $j = 1$ the Gini index

$$G(1) = \frac{2}{3} * (1 - \frac{2}{3}) + \frac{1}{3} * (1 - \frac{1}{3}) = \frac{4}{9}.$$

For $x \geq 20$, $j = 2$ the Gini index

$$G(2) = \frac{0}{200} * (1 - \frac{0}{200}) + \frac{200}{200} * (1 - \frac{200}{200}) = 0.$$

The overall Gini index at splitting $x = 20$ reads

$$Overall\ Gini = \frac{4}{9} * \frac{600}{800} + 0 * \frac{200}{800} = 1/3.$$

Since $\frac{3}{8} > \frac{1}{3}$, lower overall Gini index indicates a higher node purity, the splitting point at $x = 20$ is selected.

Case-3: Using Entropy as the measure of node homogeneity
At splitting point $x = 10$, we have

- when $x < 10$, the corresponding outcomes include A -300 and B-100, the majority vote classifies the node as Category-A. $\hat{p}_{1A} = 300/400$.

- when $x \geq 10$, the corresponding outcomes include A -100 and B-300, the majority vote classifies the node as Category-B. $\hat{p}_{2B} = 300/400$.

For $x < 10$, $j = 1$ the entropy reads

$$D(1) = -\frac{3}{4} * log(\frac{3}{4}) - \frac{1}{4} * log(\frac{1}{4}) = 0.24922.$$

For $x \geq 10$, $j = 2$ the corresponding entropy reads,

$$D(2) = -\frac{1}{4} * log(\frac{1}{4}) - \frac{3}{4} * log(\frac{3}{4}) = 0.24922.$$

The overall entropy at splitting $x = 10$ reads

$$Overall\ entropy = 0.24922 * \frac{400}{800} + 0.24922 * \frac{400}{800} = 0.24922.$$

At splitting point $x = 20$, we have

- when $x < 20$, the corresponding outcomes include A - 400 and B -200, the majority vote classifies the node as Category-A. $\hat{p}_{1A} = 400/600$.

- when $x \geq 20$, the corresponding outcomes include A - none and B -200, the majority vote classifies the node as Category-B. $\hat{p}_{2B} = 200/200 = 1$.

For $x < 20$, $j = 1$ the entropy reads

$$D(1) = -\frac{2}{3} * log(\frac{2}{3}) - \frac{1}{3} * log(\frac{1}{3}) = 0.27646.$$

For $x \geq 20$, $j = 2$ the entropy becomes

$$D(2) = 0 - \frac{200}{200} * log(\frac{200}{200}) = 0.$$

The overall entropy at splitting $x = 20$ reads

$$Overall\ entropy = 0.27646 * \frac{600}{800} + 0 * \frac{200}{800} = 0.20735.$$

Since $0.24922 > 0.20735$, lower overall entropy indicates higher node class homogeneity, and the splitting point at $x = 20$ is selected.

Note that, while the overall classification errors are equal for the two splitting points in this example, the Gini index and the entropy are able to distinguish the difference between the two. It should be emphasized that the three homogeneity measures address three different aspects in the binary splitting. In terms of the majority rule, the two splitting points $x = 10$ and $x = 20$ have equal homogeneity measures. In terms of node purity and the information conveyed, $x = 20$ is a better choice when the Gini index or entropy is used.

9.2.4 UMVUE for homogeneity in classification trees

Among the three measures of homogeneity in the construction of classification trees, the misclassification rate is essentially the probability of success of a Bernoulli random variable. Thus, the UMVUE of the misclassification rate is simply the sample proportion of the observations falling in the partition R_i. However, the UMVUE for the Gini index is a different story. Although the plug-in estimate of the Gini index gives

$$\hat{G} = \sum_k \hat{p}_{jk}(1 - \hat{p}_{jk}),$$

for each region j, it should be noted that the estimate in the above equation is not the uniformly minimum variance unbiased estimator (UMVUE). To see this point, notice that for the sample proportion $\hat{p}$ with n observations in the node,

$$E(\hat{p}(1 - \hat{p})) = E(\hat{p}) - E(\hat{p}^2)$$
$$= p - (Var(\hat{p}) + (E(\hat{p}))^2)$$
$$= p - \frac{p(1 - p)}{n} - p^2$$
$$= p(1 - p)(1 - \frac{1}{n})$$
$$= p(1 - p)\frac{n - 1}{n},$$

which is not unbiased for $p(1 - p)$.

In this regard, we have the following theorem for the UMVUE of the Gini index.

Theorem 9.1 *Assume that there are n_j observations in the branch j, $j = 1, 2$ for a binary splitting. Let X_{ijk}, $i = 1, 2, ..., n_j$ denote the event that observation i is in the class k, so that $\sum_{i=1}^{n_j} X_{ijk}$ is the total number of observations belonging to class k in the region j. The UMVUE of the Gini index in region j reads*

$$\hat{G}(j) = \sum_{k=1}^{K} \frac{\sum_{i=1}^{n_j} X_{ijk}}{n_j} (1 - \frac{\sum_{i=1}^{n_j} X_{ijk} - 1}{n_j - 1}).$$

Proof: Consider the random variable $T = \sum_{i=1}^{n_j} X_{ijk}$ in each node. Since X_{ijk} is a Bernoulli random variable with probability p_{jk}, T follows a binomial model, which is a complete distribution family. Furthermore, the sample mean T is also a sufficient statistics for p_{jk}. And T/n_j is the UMVUE of p_{jk}. What we need to find is the UMVUE of p_{jk}^2.

For any two random variables in the node, notice that

$$E(X_{1jk}X_{2jk}) = p_{jk}^2$$

we have, by the Rao-Blackwell theorem, that the UMVUE of p_{jk}^2 is

$$E(X_{1jk}X_{2jk}|T) = P(X_{1jk} = 1, X_{2jk} = 1|T)$$
$$= P(X_{1jk} = 1, X_{2jk} = 1 \ and \ T = t)/P(T = t)$$
$$= \binom{n_j - 2}{t - 2} / \binom{n_j}{t}.$$

Thus, the UMVUE of the Gini index in the region-j is

$$\hat{G}(j) = \sum_{k=1}^{K} \frac{\sum_{i=1}^{n_j} X_{ijk}}{n_j} (1 - \frac{\sum_{i=1}^{n_j} X_{ijk} - 1}{n_j - 1}).$$

9.3 Extending regression trees to range regression

As discussed in the previous sections, the decision tree uses the method of recursive binary splitting to create a partition of the feature space. It is convenient for interpretation and programming applications. However, there is a possibility that the optimal partition may not be obtained via binary splitting. For instance, other combinations of explanatory features may be able to create a different partition of the feature space that provides a better prediction on the response variable using the explanatory features. One way to explore this direction is the method of range regression, which directly improves the method of binary splitting in regression trees.

Definition 9.4 *Consider a prediction problem with response variable Y and predictor $\mathbf{x} \in S$. Let R_1, .., R_k be a partition of the feature space with optimization of the homogeneity measure. A range regression is a relationship that divides the feature space into parallel groups with homogeneity measures,*

$$\hat{f}(X) = \sum_{i=1}^{k} c_i I(X, R_i),$$

where c_i is the common mean outcome for $\mathbf{x}$ with features in the partition R_i.

Notice that, instead of making recursive binary splittings, range regression uses ranges as partitions of the feature space to establish the relationship between the outcome Y and the feature variable $\mathbf{x}$. When $R_1, ..., R_K$ are selected as the ranges of the predictor, and c_i is the mean response to the corresponding range, the overall regression method of choice is range regression.

Range regression is an analytic approach aiming to separate the source of data variation. When the original data cloud appears to contain large variations, averaging out the within group (range) variation makes the association between the response and predictor variables observable. To visually depict this point, we use an illustrating example of the relationship between residual thrombus and post-thrombotic classification of venous diseases (see, for example, [37], [35], and [24]).

The study focuses on the quantity of clot lysed in IFDVT female patients correlates with their quality of life after the operation. Equivalently, the problem emerges when evaluating the association between the CEAP score (a measurement of clinical outcomes) and residual thrombus (quantitative thrombolysis).

As shown in the plot in Figure 9.2, the conventional approach of linear regression fails to catch the association between the two variables. This is partly caused by sample fluctuations and unexpected variabilities in the patient population. The female CEAP scores vary greatly across different levels

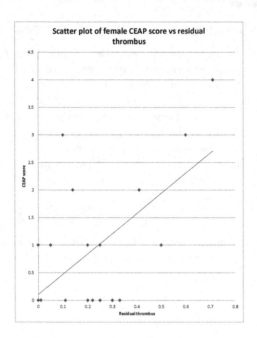

FIGURE 9.2
Post-thrombotic syndrome and Linear regression

of residual thrombus. The diagram indicates that the two random sources affect the outcome variable. It can be formulated as

$$Y_i = \alpha + \beta X_i + \epsilon_i + \eta_i,$$

where Y denotes the CEAP score, X the residual thrombus. ϵ_i and η_i denote the two different sources of random fluctuations for patient i. It should be noted that ϵ_i is for the normal fluctuation of the patients associated with the mean CEAP scores. However, η_i represents all the random sources associated with each range of residual thrombus. Obviously, the diagram shows that the change of the outcome variable (CEAP score) can not be not adequately explained by residual thrombus.

Figure 9.2 indicates that at the same level of residual thrombus, some patients have higher CEAP scores while others have relatively lower CEAP scores. The trend between the two variables of interest (CEAP scores and residual thrombus) is not observable.

To focus on subject variability for patients with similar amounts of clot lysed, Figure 9.3 shows that through range regression, stratifying patients

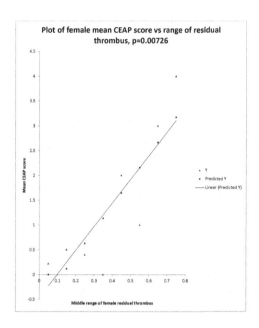

FIGURE 9.3
Post-thrombotic syndrome and Range regression

with similar amounts of clot lysed into one group identifies a measure that bundles subject variability within each stratum. The linear pattern emerges.

Similar to a regression tree, a range regression can be interpreted as follows,

$$\hat{f}(X) = \sum_{i=1}^{k} c_i I(X, R_i).$$

When a range of 10% clot lysed is set as a criterion to stratify patients, it serves as the partition of the feature space on the ranges of clot lysed. Because patients in the same stratum have similar quantities (10%) of thrombus removal, range regression averages out the impact of data variability due to confounding factors associated with each range of clot lysed. This is the homogeneity class as discussed in Section 9.1. The individual variability is then bundled with other patients in the same percentage range to show the association between clot lysed and CEAP score.

After stratifying, the sample mean of all CEAP scores for patients at the same terminal node (which contains patients having similar amounts of clot lysed) is computed as an indicator for the clinical outcome. For example, denote Y = mean female CEAP score, X = middle range of residual thrombus, ϵ = random effect, we have a fitted regression line

$$Y = -0.646 + 5.09X + \epsilon,$$

for the female patients in the data set. Since the sample mean is an asymptotically unbiased estimator of the population mean, range regression essentially models the conditional expected value of CEAP scores for each fixed range of residual thrombus as a linear trend of the residual thrombus.

$$E(\text{CEAP}|\text{residual thrombus}) = \alpha + \beta(\text{range of residual thrombus}) + \epsilon,$$

where the effect of random source η is implied by ranging the residual thrombus and averaging out the output variable within each range of residual thrombus. As shown in Figure 9.3, the method of range regression successfully reveals the association between the mean response of CEAP scores and the 10% range of residual thrombus.

At this point, we shall show a theoretical result associated with an asymptotic distribution governing the method of range regressions. It can also be analogically applied to analyze the error term in each terminal node for regression trees.

For any random vector $(x_i \quad y_i)'$, $i = 0, 1, \ldots, n$, consider the scenario where x_i falls into one of the k discernible categories $T_1, \ldots, T_k$ measured by values $s_1, \ldots, s_k$. For each x_i, there exists a set T_j such that $x_i \in T_j$, where T_j is represented by s_j,

$$s_j = \sum_{x_i \in T_j} \frac{x_i}{\#(T_j)},$$

the average of x_i in Category T_j. Then we have a mapping g(.) such that $g(x_i) = s_j, i = 1, \ldots, n, j = 1, \ldots, k$.

Let $x_{j1}, \ldots, x_{jn_j}$ be the x values corresponding to $s_j, j = 1, \ldots, k$, and denote

$$z_j = \frac{\sum_{l=1}^{n_j} y_{jl}}{n_j},$$

where $y_{j1}, \ldots, y_{jn_j}$ are the n_j response values associated with $x_{j1}, \ldots, x_{jn_j}$.

Theorem 9.2 *In the jth range, if there is a linear relationship between X and Y,*

$$Y = \alpha + \beta X + \epsilon,$$

where ϵ follows any distribution, then there exists α_1, β_1, and ϵ_1 so that

$$Z_j = \alpha_1 + \beta_1 S_j + \epsilon_1, \quad j = 1, ..., k,$$

and ϵ_1 asymptotically follows a normal model.

Proof: Consider $f_j(t)$ the moment generating function of the centralized responses $Y_{j1} - \mu_j, \ldots, Y_{jn_j} - \mu_j$. Notice that

$$f_j(t) = E[e^{t(Y_{jk}-\mu_j)}], \quad k = 1, \ldots, n_j \tag{9.1}$$
$$f_j(0) = E(1) = 1$$
$$f_j'(0) = E[(Y_{jk} - \mu_j) \times e^{0 \times (Y_{jk}-\mu_j)}] = 0$$
$$f_j''(0) = E[(Y_{jk} - \mu_j)^2 \times e^{0 \times (Y_{jk}-\mu_j)}] = \sigma_j^2.$$

Since the first two moments exist, using the Taylor expansion, we have

$$f_j(t) = 1 - \frac{1}{2}\sigma_j^2 t^2 + O(t^2). \tag{9.2}$$

The above equation implies that the moment generating function of the standardized range mean variable

$$\bar{Y}_j = \sum_{k=1}^{n_j} \frac{(Y_{jk} - \mu_j)}{\sqrt{n_j} \times \sigma_j},$$

reads

$$
\begin{aligned}
f_{n_j}(t) &= E[e^{t\bar{Y}_j}] \\
&= E\left[e^{\sum_{k=1}^{n_j} \frac{t(y_{jk}-\mu_j)}{\sqrt{n_j} \times \sigma_j}}\right] \\
&= \left[f_j\left(\frac{t}{\sqrt{n_j} \times \sigma_j}\right)\right]^{n_j}, \quad \text{by independence} \\
&= \left[1 - \frac{1}{2}\sigma_j^2\left(\frac{t^2}{n_j\sigma_j^2}\right) + O\left(\frac{t^2}{n_j\sigma_j^2}\right)\right]^{n_j} \\
&= \left[1 - \frac{1}{2}\left(\frac{t^2}{n_j}\right) + O\left(\frac{t^2}{n_j\sigma_j^2}\right)\right]^{n_j} \\
&= \left\{1 - \frac{1}{n_j}\left[\frac{t^2}{2} + n_j O\left(\frac{t^2}{n_j\sigma_j^2}\right)\right]\right\}^{n_j}.
\end{aligned}
$$

Now

$$O\left(\frac{t^2}{n_j\sigma_j^2}\right) = f_j^{(3)}(\theta)\left(\frac{t^2}{n_j\sigma_j^2}\right), \theta \in [0, \, \varepsilon_j].$$

Notice that $f_j^{(3)}(\theta)$ is bounded since it is a continuous function in a closed interval, we have, for each t,

$$n_j O\left(\frac{t^2}{n_j\sigma_j^2}\right) = f_j^{(3)}(\theta) \times \frac{t^6}{n_j^6\sigma_j^2} \longrightarrow 0 \quad \text{as} \quad n_j \longrightarrow \infty.$$

Therefore,

$$f_{n_j}(t) \longrightarrow e^{-t^2/2},$$

since

$$\lim_{n \to \infty} (1 - a/n)^n = e^{-a},$$

which is the moment generating function of $N(0, 1)$.

Thus, we have the standardized range mean variable

$$\bar{Y}_j \xrightarrow[n_j \to \infty]{} N(0, 1),$$

which means that the unstandardized range mean

$$Z_j = \frac{\sum_{l=1}^{n_j} Y_{jl}}{n_j} \xrightarrow[n_j \to \infty]{} N(\mu_j, \sigma_j/\sqrt{n_j}).$$

The second result follows from the fact that

$$\bar{Y}_j = \frac{Z_j - \mu_j}{\sigma_j/\sqrt{n_j}},$$

and

$$Z_j = \frac{\sigma}{\sqrt{n_j}}\bar{Y}_j + \mu_j.$$

Now, denote $\mathbf{z} \in R^k$, the k-dimensional vector $\mathbf{z}$ can be expressed as

$$\mathbf{z} = \boldsymbol{\mu} + \boldsymbol{\varepsilon},$$
$$\boldsymbol{\mu} = (\mu_1, \ldots, \mu_k)', \quad \boldsymbol{\varepsilon} \sim N_k(\mathbf{0}, \boldsymbol{\Sigma}),$$

where $\boldsymbol{\Sigma}$ is a $k \times k$ matrix with $\sigma_1/\sqrt{n_1}, \ldots, \sigma_k/\sqrt{n_k}$ on diagonal and 0 off-diagonals.

If $x_{ji} = s_j, i = 1, \ldots, n_j$, and suppose there is a linear relationship between μ_j and s_j, say, $\mu_j = a + bs_j$, then

$$Z_j = a + bS_j + \varepsilon_j, j = 1, \ldots, k,$$

represents a linear relationship between s_j and z_j with normal error ε_j, $j = 1, \ldots, k$.

If, on the other hand, $x_{ji} \neq s_j$, then

$$s_j \xrightarrow[n_j \to \infty]{} E(X_{ji}) = \eta_j.$$

If there is a linear relationship between μ_j and η_j, say, $\mu_j = a + b\eta_j$, then we have

$$Z_j = a + b(S_j + \zeta_j) + \varepsilon_j = a + bS_j + \zeta_j^* + \varepsilon_j,$$

which again represents a linear relationship between S_j and Z_j with a normal error ε_j, $j = 1, \ldots, k$. This concludes the proof of the theorem. More details on this theorem can be found in Kerns and Chen (2016, [75]).

The asymptotic normality of range regression model makes it legitimate in the application of linear regression analysis. Another significant advantage of range regression is the improvement of the sample correlation coefficient, as shown in the following theorem. This fits well with intuition because once the variation within each range is removed by taking the average, a stronger linear pattern shows up on the increase of the sample correlation coefficient.

Theorem 9.3 *If the X variable in each range can be represented by X_i and all ranges share the same sample size, then the sample correlation of the range regression data is larger than the sample correlation of the original data.*

Proof: Consider the original data (x_{ij}, y_{ij}), where $i = 1, \ldots, k$, $j = 1, \ldots, m_i$ and let $x_{ij} = x_i$, for all j. The sample correlation coefficient based on the original data can be calculated as

$$
\begin{aligned}
\hat{R}_1^2 &= \frac{\left(\sum_{i=1}^k \sum_{j=1}^{m_i} [(x_{ij} - \bar{x})(y_{ij} - \bar{y})]\right)^2}{[\sum_{i=1}^k \sum_{j=1}^{m_i} (x_{ij} - \bar{x})^2][\sum_{i=1}^k \sum_{j=1}^{m_i} (y_{ij} - \bar{y})^2]} \\
&= \frac{\left(\sum_{i=1}^k \sum_{j=1}^{m_i} [(x_i - \bar{x})(y_{ij} - \bar{y})]\right)^2}{[\sum_{i=1}^k \sum_{j=1}^{m_i} (x_i - \bar{x})^2][\sum_{i=1}^k \sum_{j=1}^{m_i} (y_{ij} - \bar{y})^2]},
\end{aligned} \tag{9.3}
$$

since each range has one representative value x_i for $i = 1, \ldots, k$. Now notice that

$$
\sum_{j=1}^{m_i} (y_{ij} - \bar{y}) = m_i(\bar{y}_i - \bar{y}),
$$

we have

$$
\frac{\left(\sum_{i=1}^k [(x_i - \bar{x}) \sum_{j=1}^{m_i} (y_{ij} - \bar{y})]\right)^2}{[\sum_{i=1}^k m_i(x_i - \bar{x})^2][\sum_{i=1}^k \sum_{j=1}^{m_i} (y_{ij} - \bar{y})^2]}
$$

$$
= \frac{\left(\sum_{i=1}^k [(x_i - \bar{x}) m_i(\bar{y}_i - \bar{y})]\right)^2}{[\sum_{i=1}^k m_i(x_i - \bar{x})^2][\sum_{i=1}^k \sum_{j=1}^{m_i} (y_{ij} - \bar{y})^2]},
$$

where

$$
\bar{y}_i = \frac{1}{m_i} \sum_{j=1}^{m_i} y_{ij}.
$$

Since each range contains the same amount of sample $m_i = m$, we have

$$\sum_{i=1}^{k}\sum_{j=1}^{m}(y_{ij} - \bar{y})^2 = \sum_{i=1}^{k}\sum_{j=1}^{m}(y_{ij} - \bar{y}_i + \bar{y}_i - \bar{y})^2$$

$$= \sum_{i=1}^{k}\left[\sum_{j=1}^{m}(y_{ij} - \bar{y}_i)^2 + m(\bar{y}_i - \bar{y})^2\right]$$

$$= \sum_{i=1}^{k}\sum_{j=1}^{m}(y_{ij} - \bar{y}_i)^2 + m\sum_{i=1}^{k}(\bar{y}_i - \bar{y})^2$$

$$> m\sum_{i=1}^{k}(\bar{y}_i - \bar{y})^2.$$

Now consider the transformed data $(x_i, \bar{y}_i)$, where $i = 1, \ldots, k$ and $\bar{y}_i$ is defined as the same as before. Define the overall average across the k ranges,

$$\bar{x}^* = \frac{1}{k}\sum_{i=1}^{k}x_i,$$

$$\bar{y}^* = \frac{1}{k}\sum_{i=1}^{k}\bar{y}_i = \frac{1}{k}\sum_{i=1}^{k}\left[\frac{1}{m_i}\sum_{j=1}^{m_i}y_{ij}\right].$$

The sample correlation coefficient based on the transformed data can be calculated as

$$\hat{R}_2^2 = \frac{\left(\sum_{i=1}^{k}[(x_i - \bar{x}^*)(\bar{y}_i - \bar{y}^*)]\right)^2}{[\sum_{i=1}^{k}(x_i - \bar{x}^*)^2][\sum_{i=1}^{k}(\bar{y}_i - \bar{y}^*)^2]}. \tag{9.4}$$

Notice that the data are balanced, that is, $m_i = m$, for all $i = 1, \ldots, k$, then

$$\bar{x} = \frac{1}{\sum_{i=1}^{k}m_i}\sum_{i=1}^{k}\sum_{j=1}^{m_i}x_{ij}$$

$$= \frac{1}{\sum_{i=1}^{k}m_i}\sum_{i=1}^{k}[m_i x_i]$$

$$= \frac{m\sum_{i=1}^{k}x_i}{km}$$

$$= \bar{x}^*,$$

and similarly,

$$\bar{y} = \frac{1}{\sum_{i=1}^{k} m_i} \sum_{i=1}^{k} \sum_{j=1}^{m_i} y_{ij}$$

$$= \frac{1}{mk} \sum_{i=1}^{k} [m_i \bar{y}_i]$$

$$= \frac{1}{k} \sum_{i=1}^{k} \bar{y}_i$$

$$= \bar{y}^*.$$

Equation (9.3) can be simplified further as

$$\hat{R}_1^2 = \frac{\left(\sum_{i=1}^{k} [(x_i - \bar{x})(\bar{y}_i - \bar{y})] \right)^2}{[\sum_{i=1}^{k} (x_i - \bar{x})^2][\sum_{i=1}^{k} \sum_{j=1}^{m} (y_{ij} - \bar{y})^2]}$$

$$= \frac{\left(\sum_{i=1}^{k} [(x_i - \bar{x}^*)(\bar{y}_i - \bar{y}^*)] \right)^2}{[\sum_{i=1}^{k} (x_i - \bar{x}^*)^2][\sum_{i=1}^{k} \sum_{j=1}^{m} (y_{ij} - \bar{y}^*)^2]}$$

Notice that the only difference between the above equation and Equation (9.4) is the term $\sum_{i=1}^{k} (\bar{y}_i - \bar{y}^*)^2$ in Equation (9.4). Since, $m \geq 1$

$$\sum_{i=1}^{k} \sum_{j=1}^{m} (y_{ij} - \bar{y})^2 \geq \sum_{i=1}^{k} (\bar{y}_i - \bar{y}^*)^2,$$

we have

$$\hat{R}_2^2 \geq \hat{R}_1^2.$$

This completes the proof of the theorem.

SUMMARY Besides the method of support vector machine discussed in the previous chapter for predictions without making specific model assumptions, this chapter focuses on the method of regression trees for continuous responses and classification trees for discrete responses. Additionally, we extend the method of regression trees to the method of range regression.

The highlight of this chapter concentrates on the analysis of UMVUE for the construction of decision trees. By providing a mathematical formulation for a decision tree, we present the best estimator for the homogeneity measure in the regression tree. We also include an example showing that the sample mean for responses in each terminal node is not always the best one, when the distribution of the data is not complete.

For the Gini index in classification trees, we show that the conventional estimation of the Gini index is not the best, and derive the uniformly minimum variance unbiased estimator of the population Gini index. Although, when the sample size is large, the conventional estimation on the plug-in estimation will converge to the true Gini index by the consistency property, we do not always have an infinite amount of data for slow converging predictions. This necessitates the use of best estimation for the Gini index measuring homogeneity of observations in each terminal node. The interpretation of entropy in the construction of classification tree is also addressed with examples in this chapter. Further information on trees and bagging can be found in papers such as [6], [7], [130], among others.

Range regression uses parallel splitting on ranges to replace binary splitting in regression trees. The asymptotic distribution of the range regression model bridges the data-driven camp with the model-based camp. It unifies the two data science cultures via distribution convergences of response variable sample mean.

10

Unsupervised Learning and Optimization

The previous chapters discuss data analytic issues on input features (such as predictors) relating to output features (such as the response variable), where each observation has a response. For instance, in the analysis of clinical factors related to systolic blood pressure, the response variable is the reading of patients' systolic blood pressure; in the classification of up or down market trend in the coming time period, the response variable is either "bull market" or "bear market". The model learned from the training data has a response variable intended to "supervise" the learning process by using a MSE criterion or the total probability of correct classification. However, in some data analytic problems, the response to "supervise" the learning process might not even exist. For example, in business analysis, clustering of consumer preferences helps structure the design of marketing strategies of advertising campaigns. In clinical trials and epidemiology, grouping patient symptoms helps diagnosis and prevention in public interventions. In geology, grouping on element characteristics of rock samples helps identify the main characteristic of the environment it was found. The common theme among the above mentioned applications is the lack of a response variable, due to the absence of knowledge in the experiment stage. In this chapter, we will focus on two main methods: K-means clustering and the method of principal component analysis. To briefly summarize, K-means clustering and principal component analysis are two optimization approaches in grouping a set of data.

10.1 K-means Clustering

K-mean clustering is applied when n observations is to be grouped into k ($k < n$) clusters based on feature closeness, the criterion of minimizing the total point-wise "distance". Depending on the definition of "distance", K-means clustering can be applied to various fields to quantify different types of closeness. There are a lot of publications regarding the theory and applications of K-means clustering. It is out of the scope of this book to detail the history of K-mean clustering. Early work in this field includes Forgy (1965, [49]), Hartigan and Wong (1979, [55]), and MacQueen (1967, citemacqueen1967), among others.

Definition 10.1 *Let A be a set of interest, if there exists a set of sets B_1, ..., B_t satisfying*

$$B_i \bigcap B_j = \emptyset \ for \ 1 \leq i < j \leq t$$

and

$$\bigcup_{j=1}^{t} B_j = A,$$

The the set of sets $\{B_i, ..., B_t\}$ is a partition of the set A.

In other words, the partition of a set A divides set A into finite mutually exclusive sets.

10.1.1   Clustering with Squared Euclidean Distance

Let C_1, ..., C_K be a partition of the index set $\{1, ..., n\}$. The purpose of K-means clustering is to find the partition of the p dimensional points $x_1, ..., x_n$ where $x_i \in R^p$, $i = 1, ..., n$, for the following optimization problem,

$$\hat{C}_i = \underset{C_1,...,C_K}{argmin} \{ \sum_{k=1}^{K} W(C_k) \}, \tag{10.1}$$

where, denote $|C_k|$ the number of observations in the set C_k,

$$W(C_k) = \frac{1}{|C_k|} \sum_{i,i' \in C_k} \sum_{j=1}^{p} (x_{ij} - x_{i'j})^2 \tag{10.2}$$

Notice that the definition of K-means clustering in (10.1) is defined with the squared Euclidean distance as the measurement for closeness. Such measurement can be redefined accordingly to suit different clustering problems in practice. We use the following two simple examples to clarify the terms mentioned in the definition (10.1).

Example 10.1 *When $n = 4$ and $K = 2$, find all candidate partitions of the observations.*

Solution: All the candidate partitions of the set $\{1, 2, 3, 4\}$ reads,

$$C_1 = \{1, 2\} \quad C_2 = \{3, 4\}; \qquad C_1 = \{1, 3\} \quad C_2 = \{2, 4\};$$
$$C_1 = \{1, 4\} \quad C_2 = \{2, 3\}; \qquad C_1 = \{1\} \quad C_2 = \{2, 3, 4\};$$
$$C_1 = \{2\} \quad C_2 = \{1, 3, 4\}; \qquad C_1 = \{3\} \quad C_2 = \{1, 2, 4\}$$
$$C_1 = \{4\} \quad C_2 = \{1, 2, 3\}.$$

Example 10.2 *Assume that the four points are* (11, 12, 13), (21, 22, 23), (31, 32, 33), *and* (41, 42, 43) *to be grouped into two clusters. Since* $p = 3$, *for* $C_1 = \{1, 2\}$ *and* $C_2 = \{3, 4\}$, *find the corresponding risk for optimization.*

Solution: According to Equation (10.1),

$$\sum_{k=1}^{K} [\frac{1}{|C_k|} \sum_{i,i' \in C_k} \sum_{j=1}^{p} (x_{ij} - x_{i'j})^2]$$

$$= \frac{1}{2}[(11 - 21)^2 + (12 - 22)^2 + (13 - 23)^2 + (21 - 11)^2 + (22 - 12)^2$$

$$+ (23 - 13)^2] + \frac{1}{2}[(31 - 41)^2 + (32 - 42)^2 + (33 - 43)^2 + (41 - 31)^2$$

$$+ (42 - 32)^2 + (43 - 33)^2].$$

For the partition $C_1 = \{1\}$, $C_2 = \{2, 3, 4\}$, the risk for optimization is

$$\frac{1}{3}[(21 - 31)^2 + (22 - 32)^2 + (23 - 33)^2 + (21 - 41)^2 + (22 - 42)^2 +$$

$$+ (23 - 43)^2 + (31 - 21)^2 + (32 - 22)^2 + (33 - 23)^2 + (41 - 21)^2$$

$$+ (42 - 22)^2 + (43 - 23)^2 + (41 - 31)^2 + (42 - 32)^2 + (43 - 33)^2].$$

The distance for optimization in the K-means procedure is defined as point-wise squared Euclidean distance within each cluster. If the centroid of the cluster k is defined as $(\bar{x}_{k1}, ..., \bar{x}_{kp})$ with

$$\bar{x}_{kj} = \sum_{i=1}^{|C_k|} x_{ij} = \sum_{i \in C_k} x_{ij}, \tag{10.3}$$

for each component $j = 1, ..., p$. Namely, each component of the centroid is the average of all the points in cluster C_k.

Given the definition of a centroid above, we have the following theorem transferring the sum of all point-wise distances within a cluster to the sum of all distances of points to the centroid in a cluster.

Theorem 10.1 *The mean of all point-wise distances within a cluster is twice the sum of all distances of points to the centroid of the cluster.*

Proof: Notice that the mean of all point-wise distance within a cluster reads,

$$\frac{1}{|C_k|} \sum_{i,i' \in C_k} \sum_{j=1}^{p} (x_{ij} - x_{i'j})^2,$$

and the sum of all distances of points to the centroid is

$$\sum_{i\in C_k}\sum_{j=1}^{p}(x_{ij}-\bar{x}_{kj})^2.$$

For Theorem 10.1, it suffices to prove that

$$\frac{1}{|C_k|}\sum_{i,i'\in C_k}\sum_{j=1}^{p}(x_{ij}-x_{i'j})^2 = 2\sum_{i\in C_k}\sum_{j=1}^{p}(x_{ij}-\bar{x}_{kj})^2. \qquad (10.4)$$

We start with the left-hand side of the equation (10.4).

$$\frac{1}{|C_k|}\sum_{i,i'\in C_k}\sum_{j=1}^{p}(x_{ij}-x_{i'j})^2$$

$$= \frac{1}{|C_k|}\sum_{i,i'\in C_k}\sum_{j=1}^{p}(x_{ij}-\bar{x}_{kj}+\bar{x}_{kj}-x_{i'j})^2$$

$$= \frac{1}{|C_k|}\sum_{i,i'\in C_k}\sum_{j=1}^{p}[(x_{ij}-\bar{x}_{kj})^2 - 2(x_{ij}-\bar{x}_{kj})(x_{i'j}-\bar{x}_{kj}) + (x_{i'j}-\bar{x}_{kj})^2$$

$$= \frac{|C_k|}{|C_k|}\sum_{i\in C_k}\sum_{j=1}^{p}[(x_{ij}-\bar{x}_{kj})^2 + \frac{|C_k|}{|C_k|}\sum_{i'\in C_k}\sum_{j=1}^{p}[(x_{i'j}-\bar{x}_{kj})^2$$

$$- \frac{2}{|C_k|}\sum_{i,i'\in C_k}\sum_{j=1}^{p}(x_{ij}-\bar{x}_{kj})(x_{i'j}-\bar{x}_{kj})$$

$$= 2\sum_{i\in C_k}\sum_{j=1}^{p}(x_{ij}-\bar{x}_{kj})^2,$$

since

$$\sum_{i,i'\in C_k}\sum_{j=1}^{p}(x_{ij}-\bar{x}_{kj})(x_{i'j}-\bar{x}_{kj}) = \sum_{j=1}^{p}\sum_{i\in C_k}(x_{ij}-\bar{x}_{kj})\sum_{i'\in C_k}(x_{ij}-\bar{x}_{kj}) = 0,$$

and

$$\frac{1}{|C_k|}\sum_{i,i'\in C_k}\sum_{j=1}^{p}[(x_{ij}-\bar{x}_{kj})^2 = \frac{1}{|C_k|}\sum_{i\in C_k}\sum_{i'\in C_k}\sum_{j=1}^{p}[(x_{ij}-\bar{x}_{kj})^2$$

$$= \frac{|C_k|}{|C_k|}\sum_{i\in C_k}\sum_{j=1}^{p}[(x_{ij}-\bar{x}_{kj})^2,$$

Similarly

$$\frac{1}{|C_k|}\sum_{i,i'\in C_k}\sum_{j=1}^{p}[(x_{i'j}-\bar{x}_{kj})^2 = \frac{|C_k|}{|C_k|}\sum_{i'\in C_k}\sum_{j=1}^{p}[(x_{i'j}-\bar{x}_{kj})^2.$$

The following example illustrates the main clue in the proof of Theorem 10.1.

Example 10.3 *Assume that the four points for clustering are* $(11, 12, 13)$, $(21, 22, 23)$, $(31, 32, 33)$, *and* $(41, 42, 43)$ *and the partition is* $C_1 = \{1, 2\}$, $C_2 = \{3, 4\}$. *For the cluster* C_1, *we have*

$$\frac{1}{|C_k|} \sum_{i,i' \in C_k} \sum_{j=1}^{p} (x_{ij} - x_{i'j})^2$$

$$= \frac{1}{2}[(11 - 21)^2 + (12 - 22)^2 + (13 - 23)^2 + (21 - 11)^2 + (22 - 12)^2$$
$$+ (23 - 13)^2]$$

$$= \frac{1}{2}[(11 - 16 + 16 - 21)^2 + (12 - 17 + 17 - 22)^2 + (13 - 18 + 18 - 23)^2 +$$
$$+ (21 - 16 + 16 - 11)^2 + (22 - 17 + 17 - 12)^2 + (23 - 18 + 18 - 13)^2]$$

$$= (11 - 16)^2 + (16 - 21)^2 + (12 - 17)^2 + (17 - 22)^2 + (13 - 18)^2$$
$$+ (18 - 23)^2 + (11 - 16)^2 + (16 - 21)^2 + (12 - 17)^2 + (17 - 22)^2$$
$$+ (13 - 18)^2 + (18 - 23)^2$$

$$= 2[(11 - 16)^2 + (16 - 21)^2 + (12 - 17)^2 + (17 - 22)^2 + (13 - 18)^2$$
$$+ (18 - 23)^2]$$

$$= 2[(11 - 16)^2 + (12 - 17)^2 + (13 - 18)^2 + (21 - 16)^2 + (22 - 17)^2$$
$$+ (23 - 18)^2]$$

$$= 2 \sum_{i \in C_k} \sum_{j=1}^{p} (x_{ij} - \bar{x}_{kj})^2$$

Theorem 10.1 establishes a connection from the sum of all point-wise squared distances within a cluster to the total distance from each point to the centroid of the cluster. With this theorem, it is intuitive to consider the following algorithm in the optimization process for K-means clustering.

K-means clustering algorithm

1 Initial assignment: Randomly assign a number from 1 to K to each of the observations.

2 For each cluster, compute the cluster centroid using Equation (10.3).

3 Compute the distances of each observation to the k centroids.

4 Assign each observation to the cluster that has the shortest centroid distance in the preceding step.

5 Iterate the above three steps ([2]-[4]) until the cluster assignments stop changing.

Theorem 10.2 *The K-means clustering algorithm converges to the optimal partition after finite iterations.*

Consider a partition $\Delta_t = \{C_1, ..., C_k\}$ and the total within-cluster variation,

$$f(\Delta_t) = \sum_{k=1}^{K} \frac{1}{|C_k|} \sum_{i,i' \in C_k} \sum_{j=1}^{p} (x_{ij} - x_{i'j})^2. \tag{10.5}$$

Denote

$$\hat{\Delta} = argmin_\Delta f(\Delta_t).$$

Following the K-means clustering algorithm, the minimum risk estimator $\hat{\Delta}$ exists and can be achieved after finite iterations of the algorithm.

Proof: At the first step, if $f(\Delta_1)$ is the smallest among all possible partitions, by Theorem 10.1, we have

$$\frac{1}{|C_k|} \sum_{i,i' \in C_k} \sum_{j=1}^{p} (x_{ij} - x_{i'j})^2 = 2 \sum_{i \in C_k} \sum_{j=1}^{p} (x_{ij} - \bar{x}_{kj})^2.$$

This means that the smallest possible risk is achieved at $f(\Delta_1)$, no point can be switched to a different cluster to form a different set of partition and gain any improvement on the overall within-cluster variation. Under this scenario, the MRE (minimum risk estimator) is achieved.

At any step $t, t \geq 1$, if $f(\Delta_t)$ is not the smallest value, the right-hand side of Theorem 10.1 can be improved by rearranging the points around the centroid. According to the K-means Algorithm, grouping each point to its closest centroid yields a new partition Δ_{t+1}. By Theorem 10.1, the total within-cluster variation for Δ_{t+1} is strictly less than the one for Δ_t,

$$f(\Delta_{t+1}) < f(\Delta_t). \tag{10.6}$$

Repeat the K-means algorithm with $t = 1, 2, ..., m$, we have a sequence

$$f(\Delta_1 > f(\Delta_2) > ... > f(\Delta_m > f(\Delta_{m+1}). \tag{10.7}$$

Since the distance between all the countable points are fixed, as the number of iteration m increases, the total within-cluster variation approaches the minimum value,

$$f(\Delta^*) = \sum_{k=1}^{K} \frac{1}{|C_k|} \sum_{i,i' \in C_k} \sum_{j=1}^{p} (x_{ij} - x_{i'j})^2,$$

which is Equation(10.5) in the theorem 10.2.

We shall use a numerical example to illustrate (10.6) as follows.

Example 10.4 *Assume that the four points are* $(11, 12)$, $(20, 21)$, $(40, 41)$, *and* $(51, 52)$, $K = 2$ *and*

$$\Delta_1 = \{\{(20, 21)\}, \{(11, 12), (51, 52), (40, 41)\}\}$$

we have

$$
\begin{aligned}
f(\Delta_1) &= \frac{1}{3}[(11 - 51)^2 + (12 - 52)^2 + (11 - 40)^2 + (12 - 41)^2 + \\
&\quad + (40 - 11)^2 + (41 - 12)^2 + (40 - 51)^2 + (41 - 52)^2 + \\
&\quad + (51 - 11)^2 + (52 - 12)^2 + (51 - 40)^2 + (52 - 41)^2 \\
&= 2[(11 - 34)^2 + (52 - 12)^2 + (40 - 34)^2 + (12 - 35)^2 + \\
&\quad + (41 - 35)^2 + (52 - 35)^2] \\
&= 3416.
\end{aligned}
$$

The distance of the point $(20, 21)$ *to the centroid of the first cluster* $(11, 12)$ *is smaller than its distance to the centroid of the second cluster. By rearranging the partition as*

$$\Delta_2 = \{\{(20, 21), (11, 12)\}, \quad \{(51, 52), (40, 41)\}\},$$

we have

$$f(\Delta_2) = \frac{2}{2}[(11 - 20)^2 + (12 - 21)^2 + (51 - 40)^2 + (52 - 41)^2] = 404$$

Obviously

$$f(\Delta_1) < f(\delta_2).$$

We shall discuss a numerical example to illustrate the use of the K-means clustering algorithm.

Example 10.5 *Consider a set of seven observations*

No	1	2	3	4	5	6	7
x_1	1	10	1.2	2	12	9	1.4
x_2	2	8	1.3	2.4	7	9	2

Assume that the initial random assignment sets the partition $\{C_1, C_2\}$ *as*

$$C_1 = \{1, 2, 3, 4\} \qquad C_2 = \{5, 6, 7\}.$$

Find the optimal clusters that minimizes the total within-cluster variations.

Solution: According to the K-means clustering algorithm, the first step is to find the centroids of the two clusters. The centroid of cluster-1 (x_{11}, x_{12}) reads

$$x_{11} = \frac{1}{4}(1 + 10 + 1.2 + 2) = 4.067$$

$$x_{12} = \frac{1}{4}(2 + 8 + 1.3 + 2.4) = 3.425.$$

Similarly, the centroid of C_2 reads

$$x_{21} = \frac{1}{3}(12 + 9 + 1.4) = 10.5$$

$$x_{22} = \frac{1}{3}(7 + 9 + 2) = 6.$$

Thus, we have the distance of the observations.

No	1	2	3	4	5	6	7
$d_{centroid-1}$	3.38	7.49	3.57	2.31	8.70	7.44	3.02
$d_{centroid-2}$	10.31	2.06	10.42	9.23	1.80	3.35	9.94
updated cluster	1	2	1	1	2	2	1

Now we have updated partition $\{C_1^*, C_2^*\}$ as

$$C_1^* = \{1, 3, 4, 7\} \qquad C_2^* = \{2, 5, 6\}$$

The updated centroid (x_{11}^*, x_{12}^*) becomes

$$x_{11}^* = 1.4 \quad x_{12}^* = 1.925 \qquad x_{21}^* = 10.33 \quad x_{22}^* = 8.$$

The updated point-centroid distances are

No	1	2	3	4	5	6	7
$d_{centroid-1^*}$	0.41	10.53	0.66	0.77	11.75	10.38	0.08
$d_{centroid-2^*}$	11.10	0.33	11.33	10.04	1.94	1.67	10.76
updated cluster	1	2	1	1	2	2	1

Since the observations in the updated partition are identical to the observations in the partition before checking the updated point-centroid distances, no rearrangement is needed and the optimal clusters are $\{1, 3, 4, 7\}$ and $\{2, 5, 6\}$.

The plots of the observations and the two steps in the clustering algorithm can be found in Figure 10.1.

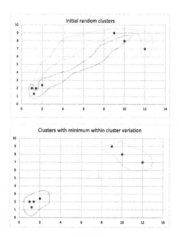

FIGURE 10.1
Numerical illustration of K-means clustering algorithm

10.1.2   Non-Euclidean Clustering

The above subsection discusses the optimizing algorithm for K-means clustering when the closeness between any two observations is measured by the squared Euclidean distance. In practice, the closeness is not always measured by the squared Euclidean distance. For example, when clustering patients into different status of diabetes based on their health features, the blood glucose level is more important than the body height. When clustering people's facial expression, index for the eyes may be of higher weight than the shape of the hair. In this scenario, some features in the dataset will have a different degree of relevance to the clustering outcome. There are different methods in clustering non-Euclidean distance. In this subsection, we will take a quick glance on non-Euclidean clustering, and use the weighted squared Euclidean distance to extend the classical K-means clustering algorithm into non-Euclidean clustering. Publications in this regard include Amorim and Mirkin (2012, [3]) as well as Amorim and Henning (2015, [2]), among others.

Let C_1, ..., C_K be a partition of the index set $\{1, ..., n\}$. Assume that we are interested in clustering the n observations in R^p into K clusters with the weights w_1, ..., w_p for the following optimization problem,

$$\hat{C}_i = \underset{C_1,...,C_K}{argmin}\{\sum_{k=1}^{K} \frac{1}{|C_k|} \sum_{i,i' \in C_k} \sum_{j=1}^{p} (x_{ij} - x_{i'j})^2\}, \qquad (10.8)$$

where, denote $|C_k|$ the number of observations in the set C_k. Notice the clustering problem in (10.8) is different from the squared Euclidean clustering (10.1) by inserting weights w_i, for $i = 1, ..., p$ where

$$0 < w_1 < 1 \quad \sum_{i=1}^{p} w_i = 1.$$

Since the K-means clustering algorithm in the preceding section is derived under the assumption that the "closeness" is measured by the squared Euclidean distance, it is inappropriate to carelessly apply the algorithm with checking the plausibility of the assumptions.

Similar to the previous subsection, we should explore the possibility of transferring the point-wise measurement to point-centroid measurement for the weighted squared Euclidean distance. In this regard, we have the following theorem.

Theorem 10.3

$$\frac{1}{|C_k|} \sum_{i,i' \in C_k} \sum_{j=1}^{p} w_j(x_{ij} - x_{i'j})^2 = 2 \sum_{i \in C_k} \sum_{j=1}^{p} w_j(x_{ij} - \bar{x}_{kj})^2. \qquad (10.9)$$

Proof Similar to the proof of Theorem 10.1, the key steps in the proof of Theorem 10.3 begin with the decomposition of the left-hand side as follows, by noticing that the weighting on the components does not affect the operation on the summation of the centroid in each cluster.

$$\frac{1}{|C_k|} \sum_{i,i' \in C_k} \sum_{j=1}^{p} w_j (x_{ij} - x_{i'j})^2$$

$$= \frac{1}{|C_k|} \sum_{j=1}^{p} w_j \sum_{i,i' \in C_k} (x_{ij} - \bar{x}_{kj} + \bar{x}_{kj} - x_{i'j})^2$$

$$= \frac{1}{|C_k|} \sum_{j=1}^{p} w_j \sum_{i,i' \in C_k} [(x_{ij} - \bar{x}_{kj})^2 - 2(x_{ij} - \bar{x}_{kj})(x_{i'j} - \bar{x}_{kj}) + (x_{i'j} - \bar{x}_{kj})^2$$

$$= \frac{|C_k|}{|C_k|} \sum_{j=1}^{p} w_j \sum_{i \in C_k} [(x_{ij} - \bar{x}_{kj})^2 + \frac{|C_k|}{|C_k|} \sum_{i' \in C_k} \sum_{j=1}^{p} [(x_{i'j} - \bar{x}_{kj})^2$$

$$\qquad - \frac{2}{|C_k|} \sum_{j=1}^{p} w_j \sum_{i,i' \in C_k} (x_{ij} - \bar{x}_{kj})(x_{i'j} - \bar{x}_{kj})$$

$$= 2 \sum_{j=1}^{p} w_j \sum_{i \in C_k} (x_{ij} - \bar{x}_{kj})^2$$

$$= 2 \sum_{i \in C_k} \sum_{j=1}^{p} w_j (x_{ij} - \bar{x}_{kj})^2.$$

With Theorem 10.3, the point-wise cluster variation with weighted squared Euclidean distances can be converted to the cluster variation from the points to the centroid of the cluster. In fact, as shown in the above proof, as long as the measurement is a linear function on the component j, converting results similar to Theorem 10.3 can be similarly proved.

The validation of Theorem 10.3, in conjunction with Theorem 10.2, consequently leads to the following non-Eculidean clustering algorithm.

Weighted squared Euclidean clustering algorithm

1 Initial assignment: Randomly assign a number from 1 to K to each of the observations.

2 For each cluster, compute the cluster centroid using Equation (10.3).

3 Compute the weighted squared Euclidean distances of each observation to the k centroids.

4 Assign each observation to the cluster that has the shortest centroid weighted distance in the preceding step.

5 Iterate the above three steps ([2]-[4]) until the cluster assignments stop changing.

10.2 Principal Component Analysis

Principal component analysis originally stems from a statistical methodology analyzing multivariate data proposed by Hotelling (1933, [66]) and Pearson (1901, [92]). It starts with the decomposition of the population covariance matrix of multivariate data into several key components constituted by the eigenvalues and eigenvectors of the population covariance matrix. When the population covariance matrix is unknown (which is the case in most practices), since the estimation of the population covariance matrix is the sample covariance matrix, the practice of principal component analysis often starts with the sample covariance matrix. Data analysts without proper statistical training usually cut corners by disregarding population covariance matrix. This partially leads to the misconception that principal component analysis begins with the sample covariance matrix, or even the standardized data. It is conceptually important to clarify this because the eigenvalues of the sample covariance matrix are not compatible due to data randomness, while the ones from the population covariance matrix are compatible because they are the parameters containing the information of the data.

10.2.1 Population Principal Components

Definition 10.2 *Let* $\mathbf{x}$ *be a p-dimensional observation from a population with covariance matrix* Σ. *Assume that the ordered eigenvalues of* Σ *are*

$$\lambda_1 \geq \lambda_2 \geq ... \geq \lambda_p \geq 0,$$

since the population covariance matrix is semi-positive, all the eigenvalues are positive. The first principal component is defined as

$$Y_1 = \mathbf{a}_1^T \mathbf{x} = a_{11} X_1 + ... + a_{1p} X_p$$

where

$$\mathbf{a}_1 = arg MAX_{\mathbf{a}_1^T \mathbf{a}=1} Var(\mathbf{a}^T \mathbf{x})$$

Similar to the definition of the first principal component in Definition 10.2, the ith principal component can be defined in a way of optimization as follows.

Definition 10.3 *Let* $\mathbf{x}$ *be a p dimension observation from a population with covariance matrix* Σ. *Assume that the ordered eigenvalues of* Σ *are*

$$\lambda_1 \geq \lambda_2 \geq ... \geq \lambda_p \geq 0.$$

The ith principal component is defined as

$$Y_i = \mathbf{a}_i^T \mathbf{x} = a_{i1} X_1 + ... + a_{ip} X_p$$

where

$$\mathbf{a}_i = arg Max_{\mathbf{a}^T \mathbf{a}=1, \, \mathbf{a}^T \mathbf{a}_j=0, \, j<i} Var(\mathbf{a}^T \mathbf{x})$$

Thus, the ith principal component of a data is essentially a linear combination of the data that is orthogonal to all the jth principal components with $j < i$.

Example 10.6 *Assume that a database contains patient information of the following features.*

$$\mathbf{x}^T = (x_1, x_2, x_3, \ldots, x_9, x_{10}),$$

where

x_1: *systolic blood pressure*
x_2: *total blood cholesterol level*
x_3: *dusty working environment*
x_4: *residential location (city, rural area)*
x_5: *transportation (car, train, bus, walk)*
x_6: *career type*
x_7: *annual income*
x_8: *medical insurance*
x_9: *heart attack/stroke history*
x_{10}: *financial investment*

We want to reduce the dimension of 10 features into several representative variables to improve the efficiency of data analysis.

Solution: Assume that the eigenvectors corresponding to the largest two eigenvalues are,

$$\mathbf{a}_1^T = (0.452, 0.603, 0, 0, 0, 0.151, -0.151, 0, 0.603, -0.151),$$

$$\mathbf{a}_2^T = (0, 0, 0.745, 0.447, 0.447, 0.149, 0.149, 0, 0, 0),$$

then the first two components become

$$Y_1 = 0.452x_1 + 0.603x_2 + 0.151x_6 - 0.151x_7 + 0.603x_9 - 0.151x_{10}$$

$$Y_2 = 0.745x_3 + 0.447x_4 + 0.447x_5 + 0.149x_6 + 0.149x_7$$

where the first component Y_1 measures blood vessel or vascular related health aspect of the patient, and the second component measures the environment aspect of the patient. In this way, the study of the 10 features of the patients can be reduced to the analysis of the two principal components measuring health and environment features of the patients. The dimension of the data is thus reduced.

The above example shows the dimensional reduction aspect of principle component analysis, but it does not show how to identify and construct the principle components. The following example shows how to construct the principle components from the covariance matrix of the data.

Example 10.7 *Suppose the random variables X_1, X_2 and X_3 have the covariance matrix*

$$\Sigma = \begin{bmatrix} 21 & 32 & 17 \\ 32 & 54 & 18 \\ 17 & 18 & 35 \end{bmatrix}$$

Find the eigenvalues and eigenvectors of the population covariance matrix.

The eigenvalues and the corresponding eigenvectors are

$$\lambda_1 = 85.25 \qquad v_1 = (-0.491, -0.754, -0.436)^T$$

$$\lambda_2 = 24.26 \qquad v_2 = (-0.055, -0.473, 0.879)^T$$

$$\lambda_3 = 0.459 \qquad v_3 = (0.869, -0.456, -0.191)^T$$

Thus, the principal components becomes

$$Y_1 = v_1^T \mathbf{x} = -0.491X_1 - 0.754X_2 - 0.436X_3$$

$$Y_2 = v_2^T \mathbf{x} = -0.055X_1 - 0.473X_2 + 0.879X_3$$

$$Y_3 = v_3^T \mathbf{x} = 0.869X_1 - 0.456X_2 - 0.191X_3$$

When we consider the largest two principal components that count 65% of the variance information in the data, the first two principal components are Y_1 and Y_2.

Evidently the above example shows how to get the principal components from the covariance matrix. In practice, the population covariance matrix is unknown. When the population covariance matrix is estimated by the sample covariance matrix, the corresponding (sample) principal components are the linear combination of the data with the corresponding eigenvectors of the sample covariance matrix. We use the following example to illustrate this point.

10.2.2 Sample principal components

As discussed in the preceding subsection, the core concept of principal component depends on the decomposition of the information or variation in the covariance matrix, which is generally unknown in the real-world practice of data analysis. However, using the consistency property of the sample covariance matrix, it is efficient to perform principal component analysis using the sample covariance matrix. It should be noted that the principal components obtained in this way are actually the sample principal components (depending on the data), not the population principal components (which does not depend on the data by definition).

Another commonly confusing issue is the standardization versus non-standardization of data analysis of principal components. One common practice is to standardize the data (shifting to the center by subtracting the sample mean, and dividing the sample standard deviation). The advantage of standardization makes the principal components invariant for location and scale transformation. However, it should be noted that performing principal component analysis on standardized data is tantamount to performing principal component analysis on the sample correlation matrix.

Theorem 10.4 *Consider a set of data x_{ij}, $i = 1, ..., k$ and $j = 1, ..., n$ for n observations of k features of a population. Denote*

$$\bar{x}_i = \frac{1}{n} \sum_j x_{ij},$$

and

$$s_{ij} = \frac{1}{n-1} \sum_{k=1}^{n} (x_{ik} - \bar{x}_i)(x_{jk} - \bar{x}_j).$$

The standardized observations

$$z_{ij} = \frac{x_{ij} - \bar{x}_i}{\sqrt{s_{ii}}},$$

where $i = 1, ..., k$, $j = 1, ..., n$, by centralizing to the sample mean and dividing by the sample standard deviation. We have

$$S_z = R_x \tag{10.10}$$

The sample covariance matrix of the standardized data is the correlation matrix of the original data.

Proof: Denote the data matrix of the standardized data by Z. Notice that for the standardized data, the sample mean of each component reads

$$\bar{z} = \frac{1}{n}(\mathbf{1}'Z)' = \frac{1}{n}Z'\mathbf{1} = 0,$$

since

$$\frac{1}{n} \sum_{j=1}^{n} \frac{x_{ji} - \bar{x}_i}{\sqrt{s_{ii}}} = \frac{1}{n\sqrt{s_{ii}}}(\sum_{j=1}^{n} x_{ji} - n\bar{x}_i) = 0.$$

Now, by the definition, the sample covariance matrix of the standardized data is

$$S_z = \frac{1}{n-1}(\sum_{k=1}^{n}(z_{ki} - \bar{z}_i)(z_{kj} - \bar{z}_j)) = \frac{1}{n-1}Z'Z.$$

Now, notice that

$$\frac{1}{n-1}Z'Z = \frac{1}{n-1}\left(\frac{(n-1)s_{ij}}{\sqrt{s_{ii}}\sqrt{s_{jj}}}\right) = R_x.$$

The meaning and interpretation of principal component analysis on the correlation matrix are different from those obtained from the sample covariance matrix. We use the following example to illustrate this point.

Example 10.8 *Consider the following hypothetical dataset for a random sample of 9 records of three stocks in percentages.*

$$stock1 = (6, 8, 9, 3, 8, 7, 9, 8, 2)$$

$$stock2 = (-2, 15, 9, 3, 7, 7, -8, -14, 7)$$

$$stock3 = (2, 3, 2, 4, 3, 3, 3, 2, 2),$$

Find the principal components of the stock market performance.

Solution: As shown in Figure 10.2, the components of the sample covariance matrix of the data.frame reads

$$Y_1 = 0.044Stock1 - 0.999Stock2 - 0.017Stock3$$

$$Y_2 = 0.999Stock1 + 0.045Stock2 - 0.032Stock3$$

$$Y_3 = 0.033Stock1 - 0.015Stock2 + 0.999Stock3,$$

with eigenvalues $\lambda_1 = 82.922$, $\lambda_2 = 6.357$, and $\lambda_3 = 0.471$. The first principal component takes

$$\frac{82.922}{82.922 + 6.357 + 0.471} = 92.4\%$$

of the total variation of the data. In fact, since Stock 3 (performing like CD or bonds) basically remains the same among the 9 sample years, and Stock1 changes slightly (performing like low-risk portfolios), the market variation is essentially reflected by Stock 2, while Stock 1 and Stock 3 carry less weights in the evaluation of the market performance.

However, with the data standardization approach, when we subtract the sample mean and divide each observation with the sample standard deviation, the components of the sample correlation matrix reads

$$Y_1^* = 0.515Stock1 - 0.609Stock2 - 0.603Stock3$$

$$Y_2^* = -0.856Stock1 - 0.332Stock2 - 0.395Stock3$$

```
> data<-data.frame(stock1=c(6, 8, 9, 3, 8, 7, 9, 8, 2),
+ stock2=c(-2, 15, 9, 3, 7, 7, -8, -14, 7),
+ stock3=c(2, 3, 2, 4, 3, 3, 3, 2, 2))

> h=cov(data)

> j=cor(data)

> eigen(h)
eigen() decomposition
$`values`
[1] 82.9223567  6.3569839  0.4706594

$vectors
              [,1]         [,2]        [,3]
[1,]  0.04416809  0.99848413  0.03284249
[2,] -0.99888288  0.04469073 -0.01535322
[3,] -0.01679770 -0.03212768  0.99934261


> eigen(j)
eigen() decomposition
$`values`
[1] 1.3345071 0.8794526 0.7860403

$vectors
              [,1]         [,2]        [,3]
[1,]  0.5150182 -0.8562234 -0.04046871
[2,] -0.6093253 -0.3324862 -0.71984420
[3,] -0.6028922 -0.3953915  0.69295498
```

FIGURE 10.2

Sample covariance matrix vs sample correlation matrix

$$Y_3^* = -0.04 Stock1 - 0.72 Stock2 + 0.693 Stock3,$$

with eigenvalues $\lambda_1 = 1.334$, $\lambda_2 = 0.879$, and $\lambda_3 = 0.786$. Thus, the first principal component of the correlation matrix carries

$$\frac{1.334}{1.334 + 0.879 + 0.786} = 44.5\%$$

of the total information/variation. To obtain more than 60% of the total variation, the first two principal components should be used. Obviously, the first component of the standardized data measures the contrast between Stock1 and the combination of Stock2 and Stock3, the second component of the standardized data focuses on Stock1 and the last one focuses on the contrast between Stock2 and Stock3.

It is not difficult to see that after standardization, the difference on the variation of the three stocks has reduced due to the scaling of the variable by its sample standard deviation. With standardization, different variations on different variables are removed or rescaled, which causes misleading conclusion in data analysis.

This example shows that when we rescale data through standardization, the variations of the original variables are shrunk to one. This causes distortion of the variation information conveyed in the original dataset.

SUMMARY Preceding chapters describe prediction problems involving a measurable response variable, a supervised measurement in the learning process. This chapter, on the other hand, discusses methods in the case where we do not have a response variable for classification or prediction, an unsupervised learning scenario. In particular, we discuss the method of K-mean clustering with squared Euclidean distance as homogeneity measurement, and extend the method into cases where the classification criterion is non-Euclidean distance. We also illustrate the method of principle component analysis addressing the difference between the population principle components and sample principle components. Recent development on the application of K-means clustering method to the impact of masking policy during Covid-19 pandemic can be found in Chen and Chen (2024, [36]), among others.

11

Simultaneous Learning and Multiplicity

This chapter discusses a common situation in data science when two or more populations are involved in the learning process. When we have only one data population for prediction, the error rates (false positive, false negative) are clearly defined. However, when two or more sets of data are involved in the analysis process, since each path of the analysis generates error rates, controlling inference error in one population (one statement) does not control the overall error rates. In fact, the overall error rates accumulate as the number of data path increases. This chapter thus discusses methods applicable to adjust the multiplicity in multi-path statistical learning, the scenario where two or more sources of datasets are involved in the prediction process.

One of the special case of multiple path learning is the case when the number of observations increases (instead of being fixed) in the learning process. To this end, we discuss the method of sequential analysis, where the two error rates (false positive and false negative) are combined to determine the required sample size. This is the method of sequential analysis where data coming at different phases.

Another approach to handling multi-resources learning simultaneously is the methodology of multiple comparisons. We shall focus on recent developments on simultaneous confidence segments for dose-response curves, and weighted hypotheses with high dimensional data in this chapter.

Materials in this chapter essentially synthesize some recent publications in simultaneous inference, including Ma et al. (2023)[82], Yu et al (2022)[128], Chen (2016)[23], and Kerns and Chen (2017)[76].

11.1 Sequential Data

One of the common questions in data analysis is the determination of sample size. Notice that in previous chapters, we focus on the control of type-I error with a fixed sample size. Consider the scenario where the type-I error and type-II error are specified, which can be treated as the false positive (incorrectly rejecting the null hypothesis) and false negative (incorrectly rejecting the alternative hypothesis) error rates. The sample size computation is available for the inference on population mean when the population standard deviation is given. However, in many applications, the population standard deviation is unknown. Under this scenario, the two-stage sequential procedure can be applied. In the literature, there are various excellent review articles on sequential analysis, such as Lai (2001, [77]). In this section, we will use examples to describe the basic idea of prediction with sequential data.

11.1.1 Wald's sequential likelihood ratio test (SPRT)

We start with some simplest cases on the application of Wald's SPRT test.

Let A and B be two constants in Wald's SPRT procedure, which satisfy the following equations,

1) P(type-I error)=P(Incorrectly rejecting the null hypothesis)=α, and

2) P(type-II error)= P(Incorrectly rejecting the alternative hypothesis)=β.

We have the following general procedure in sequential inference with the above two criteria. For testing H_0: $\mu = \mu_0$ versus H_1: $\mu = \mu_1$, denote the likelihood ratio

$$R_n(\mu_0, \mu_1) = \frac{\prod_{i=1}^{n} f(x_i|\mu_1)}{\prod_{j=1}^{n} f(x_i|\mu_0)},$$

and the likelihood test statistic as Λ_n.

- Reject the null hypothesis (or equivalently accepting the alternative hypothesis) if the data satisfies $(\Lambda_n|\mu = \mu_0) > A$; clear data evudebce for H_1.

- Reject the alternative hypothesis (or equivalently accepting the null hypothesis) if the data satisfies $(\Lambda_n|\mu = \mu_1) < B$; clear data evidence for H_0.

- Continue sampling (without making any conclusion on accepting or rejection the null hypothesis) if the data satisfies $A \geq (\Lambda_n|\mu = \mu_0)$ and $(\Lambda|\mu = \mu_1) \geq B$.

Example 11.1 *Consider a testing problem on $\mu = 2$ versus $\mu = 6$ for data drawn from a population fallowing the normal model $N(\mu, 4)$ with the variance $\sigma^2 = 4$.*

Solution If $\mu = 6$ the data contain higher likelihood for the alternative hypothesis. If $\mu = 2$, on the other hand, the data contain higher likelihood to support the null hypothesis. Thus, for sprt, denote $f(x|\mu)$ the density of the underlying model, and

$$R_n = \frac{\prod_{i=1}^{n} f(x_i|\mu = 6)}{\prod_{j=1}^{n} f(x_i|\mu = 2)},$$

the decision rule of *SPRT* can be described as follows.

Since

$$f(x|\mu) = \frac{1}{2\pi\sigma}e^{-\frac{1}{2\sigma^2}(x-\mu)^2},$$

after some simple algebra, the likelihood ratio can be simplified as

$$R_n = Exp\{-\frac{1}{2\sigma^2}\sum_{i=1}^{n}[(x_i - \mu_1)^2 - (x_i - \mu_0)^2]\}$$

$$= Exp\{\frac{1}{2\sigma^2}[2n\bar{X}(\mu_1 - \mu_0) + n(\mu_0^2 - \mu_1^2)]\}.$$

Thus,

$$R_n > A$$

is equivalent to

$$\bar{X} > c$$

for some constant c. The test statistic is $\Lambda_n = \bar{X}$, and

$$P(\Lambda_n > A|\mu = \mu_0) = \alpha$$

is tantamount to

$$\frac{1}{\sigma}(\bar{X} - \mu_0)\sqrt{n} > Z_{1-\alpha}.$$

Similarly, the Wald's sequential probability ratio test for rejecting the alternative hypothesis, in this simple setting, becomes

$$\frac{1}{\sigma}(\bar{X} - \mu_1)\sqrt{n} < Z_\beta.$$

Summarizing the above discussion on the Wald's sequential probability ratio test, we have the following decision criteria for the inference problem discussed in Example 11.1

- Reject the null hypothesis (or equivalently accepting the alternative hypothesis) if the data satisfied the condition

$$\frac{1}{\sigma}(\bar{X} - \mu_0)\sqrt{n} > Z_{1-\alpha};$$

- Reject the alternative hypothesis (or equivalently accepting the null hypothesis) if the data satisfies the condition

$$\frac{1}{\sigma}(\bar{X} - \mu_1)\sqrt{n} < Z_\beta;$$

- Continue sampling without making any conclusion, if

$$Z_\beta \leq \frac{1}{\sigma}(\bar{X} - \mu_1)\sqrt{n}, \text{ and } \frac{1}{\sigma}(\bar{X} - \mu_0)\sqrt{n} < Z_{1-\alpha}.$$

The above discussion is valid when the data follows a normal model with a given standard deviation. However, in practice, the standard deviation is unknown. This leads to the *sprt-t* package in R for Wald's sequential probability ratio test without knowing the population standard deviation.

Another practical issue related to the analysis of sequential data is the statement claiming the non-difference between the hypothesized mean and the true but unknown population mean in composite hypothesis $\mu = \mu_0$ versus $\mu \neq \mu_0$. Since the datum varies at certain level, and it makes more sense to claim closeness within certain range instead of claiming an inequality vaguely between the unknown and the hypothesized value. This is because an inequality could imply that the unknown value is very close to the hypothesized value or very far away from the hypothesized value. Toward this end, a practical concept to resolve the problem, is the effective size.

Definition 11.1 Effective Size *Assume that the hypothesized mean is μ_0, and the population standard deviation is σ. For any population mean μ, the effective size measures the distance between μ and μ_0 adjusted by the population standard deviation,*

$$d = \frac{|\mu - \mu_0|}{\sigma}.$$

With the concept of effective size, the composite hypothesis $\mu = \mu_0$ versus $\mu \neq \mu_0$ is specified, in sequential analysis, as $\mu = \mu_0$ versus $|\mu - \mu_0| > d\sigma$. The alternative hypothesis is interpreted as the true but unknown mean value is d standard deviation away from the hypothesized mean μ_0. Heuristically, when the effective size decrease, the likelihood ratio in Wald's SPRT decreases because usually it is harder to detect the mean difference when the true value and the hypothesized mean are very close. For instance, given a specified effective size d_0, since the alternative hypothesis is $|\mu - \mu_0| > d\sigma$, we have,

$$P(Type - II\ error) = \sup_{\frac{|\mu-\mu_0|}{\sigma} > d_0} P(accepting\ H_0 | \frac{|\mu - \mu_0|}{\sigma} > d_0)$$

$$= P(accepting\ H_0 | \frac{|\mu - \mu_0|}{\sigma} = d_0)$$

$$= P(accepting\ H_0 | d = d_0).$$

We shall use an example to discuss the R-package "sprit" for sequential data analytics using the sprt t-test.

Example 11.2 *For convenience, consider the income data in the R-package sqrtt. The dataset contains 120 observations on monthly income for 60 male and 60 female. For illustration purpose, we will use the income data to examine the impact of effective size on the inference outcome. We first compare the mean monthly income level of male with that of female under different effective sizes, then use the data to test the sequential information with SPRT on a* *specific value for the population mean of the monthly income, and conclude the example with a discussion regarding the impact of the alternative likelihood on effective size.*

Case 1. sprt t-test comparing two population means

Assume that we are interested in detecting the mean monthly income differences between male and female.

```
> seq_ttest(monthly_income ~ sex, data =df_income, d=0.2)

***** Sequential Two Sample t-test *****

data: monthly_income ~ sex
test statistic:
 log-likelihood ratio = -0.59447, decision = continue sampling
SPRT thresholds:
 lower log(B) = -2.94444, upper log(A) = 2.94444
Log-Likelihood of the:
 alternative hypothesis = 0.82689
 null hypothesis = 1.42137
alternative hypothesis: true difference in means is not equal to 0.
specified effect size: Cohen's d = 0.2
degrees of freedom: df = 118
sample estimates:
mean of x mean of y
 3072.086 3080.715

> seq_ttest(monthly_income ~ sex, data =df_income, d=0.8)

***** Sequential Two Sample t-test *****

data: monthly_income ~ sex
test statistic:
 log-likelihood ratio = -9.51391, decision = accept H0
SPRT thresholds:
 lower log(B) = -2.94444, upper log(A) = 2.94444
Log-Likelihood of the:
 alternative hypothesis = -8.09254
 null hypothesis = 1.42137
alternative hypothesis: true difference in means is not equal to 0.
specified effect size: Cohen's d = 0.8
degrees of freedom: df = 118
sample estimates:
mean of x mean of y
 3072.086 3080.715
```

FIGURE 11.1
Mean monthly income between male and female with *sprr*

As shown in the output Figure 11.1, when the effective size is set to 0.2, the null and alternative hypotheses become

$$H_0 : \mu_{male} = \mu_{female} \quad versus \quad H_1 : |\mu_{male} - \mu_{female}| > 0.2\sigma,$$

Namely to test whether the mean monthly income difference is more than 20% of the data variation. The SPRT thresholds are

$$log(B) = -log(0.95/0.05) = -2.9444(= -log(A))$$

since the probability of type-I error and the probability of type-II error are set to 0.05 in the coding. The log-likelihood of the null hypothesis

$$log \prod_i f(x_i | \mu_{male} = \mu_{female}) = 1.42137;$$

The log-likelihood of the alternative hypothesis

$$sup \, log \prod_i f(x_i | (|\mu_{male} - \mu_{female}|) > 0.2\sigma) = 0.82689.$$

and the log-likelihood ratio reads

$$sup \, log \prod_i f(x_i | (|\mu_{male} - \mu_{female}|) > 0.2\sigma) = 0.82689.$$

Thus the log-likelihood ratio becomes

$$log - likelihood \, ratio$$
$$= sup \, log \prod_i f(x_i | (|\mu_{male} - \mu_{female})| > 0.2\sigma)$$
$$- log \prod_i f(x_i | \mu_{male} = \mu_{female})$$
$$= 0.82689 - 1.42137$$
$$= -0.59447.$$

Since the value of the log-likelihood ratio is within the two threshold

$$-0.59447 \in (-2.9444, 2.9444),$$

there is no data evidence to reject the null hypothesis or the alternative hypothesis. According to Wald's SPRT, the inference outcome is "to continue sampling". This means that the existing sample size is not large enough to detect the mean monthly income difference within 20% of the data variation, thus the testing conclusion is " to continue sampling". In fact the two sample means of monthly income are very close to each other. The mean monthly income of male is $3072.09, and that of female is $3080.72.

It should be noted that the conclusion of "to continue sampling" is for the pre-specified effective size of 0.2. When the effective size changes, the inference conclusion changes. For instance, with the same dataset, when the effective size is changed to 0.8, the corresponding log-likelihood of the alternative space becomes

$$sup \, log \prod_i f(x_i : |\mu_{male} - \mu_{female}| > 0.8\sigma) = -8.09254.$$

and the corresponding log-likelihood ratio becomes

$$log - likelihood\ ratio$$

$$= sup\ \{log \prod_i f(x_i | (|\mu_{male} - \mu_{female}|) > 0.8\sigma)$$

$$- log \prod_i f(x_i | \mu_{male} = \mu_{female})\}$$

$$= -8.09254 - 1.42137$$

$$= -9.51391.$$

Since the value of the likelihood ratio is less than -2.9444, according to Wald's SPRT, the conclusion is to accept the null hypothesis, which means that the mean monthly income of male, on average, has no significant difference within 80% of the data variation. In this case, the data contains evidence to support the claim that the mean monthly income difference between male and female, is not beyond 80% of the data variation. This echoes with the early discussion on the relationship between the effective size and the inference conclusion.

Case 2. Wald's SPRT t-test on a hypothesized mean

Now, assume that, instead of comparing two means, we are interested in testing whether the population mean (monthly income) is within certain range of a fixed value. For instance, assume that the hypothesis is

$$\mu = 3100 \quad versus \quad |\mu - 3100| > 0.5\sigma.$$

Namely the hypothesized mean of the monthly income is set to $3100, and we want to see whether there is any data evidence supporting the claim that the difference is more than 50% of the data variation.

As shown in Figure 11.2, the log-likelihood of the null hypothesis is -0.41656, and the one for the alternative hypothesis reads -13.20851. This makes the log-likelihood ratio

$$-13.20851 - (-0.41656) = -12.79159,$$

which is less than the lower threshold -2.9444. According to Wald's SPRT, the optimal decision is to accept the null hypothesis. As a matter of fact, the sample mean, $3076.4, is indeed well within the range if 50% data variation, which supports the Wald's SPRT in accepting the null hypothesis.

For comparison purpose, when the hypothesized mean is set to $1200, as shown in the second half of Figure 11.2, the value of the hypothesized mean $1200 is far below the sample mean, the corresponding log-likelihood value for the null hypothesis is -169.7469, while the one for the alternative hypothesis is -119.9379, the log-likelihood ratio reads 49.8090, which is larger than the upper threshold of the Wald's SPRT, 2.9444. Thus the inference conclusion is to accept the alternative hypothesis. In fact, the sample mean monthly income

```
> seq_ttest(monthly_income ~ 1, mu=3100, data =df_income, d=0.5)

***** Sequential One Sample t-test *****

data: monthly_income ~ 1
test statistic:
 log-likelihood ratio = -12.79195, decision = accept H0
SPRT thresholds:
 lower log(B) = -2.94444, upper log(A) = 2.94444
Log-Likelihood of the:
 alternative hypothesis = -13.20851
 null hypothesis = -0.41656
alternative hypothesis: true mean is not equal to 3100 .
specified effect size: Cohen's d = 0.5
degrees of freedom: df = 119
sample estimates:
mean of x
  3076.4

> seq_ttest(monthly_income ~ 1, mu=1200, data =df_income, d=0.5)

***** Sequential One Sample t-test *****

data: monthly_income ~ 1
test statistic:
 log-likelihood ratio = 49.80902, decision = accept H1
SPRT thresholds:
 lower log(B) = -2.94444, upper log(A) = 2.94444
Log-Likelihood of the:
 alternative hypothesis = -119.9379
 null hypothesis = -169.7469
alternative hypothesis: true mean is not equal to 1200 .
specified effect size: Cohen's d = 0.5
degrees of freedom: df = 119
sample estimates:
mean of x
  3076.4
```

FIGURE 11.2

Overall mean monthly income vs a value with *sprtt*

of the data is \$3076.4, which is beyond the hypothesized mean (\$1200) by more than 50% of the data variation.

Case 3. Effective size and likelihood under alternative hypothesis

To further explore the relationship between the effective size and the log-likelihood of the alternative hypothesis, we plot the corresponding log likelihood values of the alternative hypothesis for every 10% increase of the effective size.

The plot in Figure 11.3 clearly shows that for this dataset, the likelihood value of the alternative space decreases as the effective size increases. This is consistent with the fact that it is harder to detect smaller difference between the true mean and the hypothesized mean.

```
Log_alternative=c()
decision=c()
Effective_Size_d=seq(0.1,.9, by=.1)
for (i in 1:9) { sstest=seq_ttest(monthly_income ~ sex, data =df_income,
d=Effective_Size_d[i])
Log_alternative[i]=sstest@likelihood_1
}
plot(Effective_Size_d, Log_alternative,type = 'l', main="Effect of Effective_Size_d
on Alternative Likelihood" )
points(Effective_Size_d, Log_ratio)
```

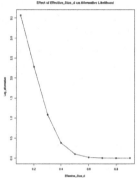

FIGURE 11.3
Log-likelihood ratio vs effective sizes

11.1.2 Two-stage Estimation for Sequential Data

In this section, we shall discuss the two-stage estimation method in the analysis of sequential data. Consider the sample size of a given set of normal data $X_1, ..., X_k$ in the process of testing

$$H_0 : \mu = \mu_0 \quad versus \quad H_1 : \mu = \mu_0 + \delta,$$

for any given value $\delta > 0$. It is well known that when the population standard deviation is given as σ, the probability of the type-II error becomes

$$P(accepting\ H_0|H_1\ true) = P(\frac{\bar{X} - \mu_0}{\sigma/\sqrt{k}} < Z_{1-\alpha}|\mu_0 + \delta)$$

$$= P(\frac{\bar{X} - \mu_1}{\sigma/\sqrt{k}} < Z_{1-\alpha} - \frac{\delta}{\sigma/\sqrt{k}}|H_1)$$

$$= P(Z < Z_{1-\alpha} - \frac{\delta}{\sigma/\sqrt{k}}),$$

also, for given probabilities of type-I and type-II errors, the minimal sample size for one-stage sampling is

$$k \geq \frac{(Z_{1-\alpha} + Z_{1-\beta})^2 \sigma^2}{\delta^2}.$$

The above analysis assumes the population standard deviation, however, in practice, the population standard deviation is unknown. Where we replace the population standard deviation with the sample standard deviation, the above discussion becomes invalid. This is because when σ is replaced by the sample standard deviation s, the latter is a random variable depending on (being a function of) the sample size k. Under this scenario, the one-stage approach is unable to solve the difficult on sample size determination. Toward this end, we need the following theorem.

Consider the testing problem $H_0 : \mu = \mu_0$ versus $H_1 : \mu = \mu_0 + \delta$, $\delta > 0$ for a normal population with unknown mean μ and unknown standard deviation σ. In a sequential sampling, assume that the first n_0 observations are available, $X_1, ..., X_{n_0}$. Denote α and β the required probabilities of type-I and type-II errors. Let s_0 be the sample standard deviation of the first sample (the first n_0 observations).

Theorem 11.1 *If a sequential sample of $n - n_0$ observations is available with*

$$n = max\{[\frac{(t_{n_0-1,1-\alpha} + t_{n_0-1,1-\beta})^2 s_0^2}{\delta^2}] + 1, n_0\},$$

denote $\bar{X}_n$ the sample mean of the updated sample, then the rejection region

$$R = \frac{\bar{X}_n - \mu_0}{s_0/\sqrt{n}} > t_{n_0-1,1-\alpha},$$

has both probabilities of type-I error and type-II error controlled at α and β levels, respectively.

Proof: . It suffices to prove the following two conditions for the theorem. The first one is for the probability of type-I error in the updated sample.

$$P(incorrectly\ rejecting\ H_0) = P(\frac{\bar{X}_n - \mu_0}{s_0/\sqrt{n}} > t_{n_0-1,1-\alpha}|\mu_0)$$

$$= P(t_{n_0-1} > t_{n_0-1,1-\alpha})$$

$$= \alpha.$$

As for the second condition on the control of the type-II error, notice that

$$n > \frac{(t_{n_0-1,1-\alpha} + t_{n_0-1,1-\beta})^2 s_0^2}{\delta^2},$$

we have

$P(\text{incorrectly accepting } H_0)$

$$= P(\frac{\bar{X}_n - \mu_0}{s_0/\sqrt{n}} < t_{n_0-1,1-\alpha}|\mu_0 + \delta)$$

$$= P(\frac{\bar{X}_n - \mu_1}{s_0/\sqrt{n}} < t_{n_0-1,1-\alpha} - \frac{\delta}{s_0/\sqrt{n}}|\mu_1)$$

$$= P(T_{n_0-1} < t_{n_0-1,1-\alpha} - \frac{\delta}{s_0}\sqrt{n})$$

$$\leq P(T_{n_0-1} < t_{n_0-1,1-\alpha} - \frac{\delta}{s_0}\sqrt{(\frac{(t_{n_0-1,1-\alpha} + t_{n_0-1,1-\beta})^2 s_0^2}{\delta^2})})$$

$$= P(T_{n_0-1} < t_{n_0-1,1-\alpha} - (t_{n_0-1,1-\alpha} + t_{n_0-1,1-\beta}))$$

$$= P(T_{n_0-1} < -t_{n_0-1,1-\beta})$$

$$= \beta.$$

The following example shows how to use Theorem 11.1 for data analysis.

Example 11.3 *The following data set contains one-week trading prices of a stock at NYSC* {14.98, 15.09, 15.12, 15.15, 15.22, 14.98, 14.96}. *Does the dataset contain enough information to test whether the mean price is $15.00 or $15.05 at 0.05 significance level with 95% power? If not, how many additional observations do we need in sequential sampling?*

Solution: When the null hypothesis is $\mu = \$15$, the corresponding t-value is 1.889, which leads to the p-value of 0.054. Since the p-value is larger than the nominated 0.05 level, we do not have data evidence to reject the null hypothesis. On the other hand, when the alternative hypothesis is assumed true, $\mu = \$15.05$, the corresponding t-value is 0.567, which results in a p-value of 0.704. The p-value is much larger than the nominated 0.05 level for the probability of making type-II error, thus, there is no data evidence to reject the alternative (or equivalently, to accept the null hypothesis) at 0.05 level.

Summarily, to control both the probabilities of type-I error and type-II error at 0.05 level, the information on the seven-day stock trading price is not enough. Based on Wald's SPRT, the conclusion is "to continue sampling" for this set of data, as shown in Figure 11.4.

To control the probability of type-I error and the probability of type-II error at 0.05 level, on testing whether the mean stock selling price is $0.05 away from the $15.00 level, we need 61 observations, as shown in Figure 11.5. With $n_0 = 7$, additional 54 stock exchange prices are needed for this question.

Theorem 11.1 used a two-stage sequential method to solve the problem on sample size requirement for inference problems when the population standard deviation is unknown. A related question is to estimate the unknown mean

```
stockdata <- c(14.98, 15.09, 15.12, 15.15, 15.22, 14.98, 14.96)
TwoStageTest <- function(x,theta0,theta1,alpha,beta){
  xbar <- mean(x)
  s <- sd(x)
  n0 <- length(x)
  tvalue0 <- (xbar-theta0)/(s/sqrt(n0))
  tvalue1 <- (xbar-theta1)/(s/sqrt(n0))

  p0 <- pt(tvalue0, df=n0-1, lower.tail = FALSE)
  p1 <- pt(tvalue1, df=n0-1, lower.tail = TRUE)
  output <- NULL
  # Decision
  if (p0 <= alpha) {
    output <- "Accept H0"
  } else if (p1 <= beta) {
    output <- "Accept H1"
  } else {
    output <- "Continue sampling"
  }
  return(list(t0=tvalue0, p0=p0,t1=tvalue1,p1=p1,output))
}
TwoStageTest(stockdata, theta0=15.000,theta1=15.05,alpha=0.05,beta=0.05)

$t0
[1] 1.888474

$p0
[1] 0.05394006

$t1
[1] 0.5665422

$p1
[1] 0.7042146

[[5]]
[1] "Continue sampling"
```

FIGURE 11.4
Codes for two-stage sequential Student-t test

with unknown variance and the requirement that the accuracy of the estimation is controlled at a pre-specified level. In fact, the idea in Theorem 11.1 can be further extended to the two-stage confidence estimation method as follows.

Definition 11.2 *For a set of data* $\mathbf{X}$, *the terminology estimation error refers to the following. If a* $(1-\alpha)\%$ *confidence interval of an unknown parameter* θ *is*

$$\hat{\theta}(\mathbf{X}) - e(\mathbf{X}) \le \theta \le \hat{\theta}(\mathbf{X}) + e(\mathbf{X}),$$

the statistic $e(\mathbf{X})$ *is the* **estimation error**.

Estimation error is a measure of evaluating the prediction accuracy in confidence interval estimation. Lower estimation error implies that the estimator is closely toward the true value, which can be interpreted as a higher accuracy level of the estimation. In confidence interval estimation with fixed sample size, usually higher confidence level leads to lower accuracy level. For a given sample size, it is impossible to control the accuracy and confidence level at the same time. However, using the method of two-stage sampling in sequential analysis, similarly to Theorem 11.1, we can select a sample size (in the second stage) that controls both confidence level and accuracy.

Consider the estimation of a population mean μ with unknown standard deviation for normal populations. Denote E the pre-specified estimation error,

```
stockdata <- c(14.98, 15.09, 15.12, 15.15, 15.22, 14.98, 14.96)

TwoStageTestSize <- function(x,theta0,theta1,alpha,beta){

 n0 <- length(x)

 s <- sd(x)

 delta <- theta1-theta0

 t1 <- qt(1-alpha, n0-1, lower.tail = FALSE)

 t2 <- qt(1-beta, n0-1, lower.tail = FALSE)

 z <- delta**2/(t1+t2)**2

 n <- ceiling(s**2/(z))

 return(n)

 }

TwoStageTestSize(stockdata,15,15.05,0.05,0.05)

[1] 61
```

FIGURE 11.5
Sample size calculation for stock data-2

and $1 - \alpha$ the confidence level. Denote n_0 the sample size of the first n_0 observations. Let s_0 be the sample standard deviation of the first sample (the first n_0 observations).

Theorem 11.2 *If there are $n - n_0$ observations available in the second sample, where*

$$n = max\{[(s_0 \frac{t_{n_0-1,1-\alpha}}{E})^2] + 1, n_0\},$$

denote $\bar{X}_n$ the sample mean of the updated sample with n observations the confidence interval

$$A = (\bar{X}_n - E, \bar{X} + E),$$

has confidence level $1 - \alpha$.

Proof: Notice that the sample size in the second stage satisfies

$$n > (s_0 \frac{t_{n_0-1,1-\alpha}}{E})^2,$$

we have

$$P(\bar{X}_n - E \leq \mu \leq \bar{X} + E)$$
$$= P(|\bar{X} - \mu| \leq E)$$
$$= P(\frac{|\bar{X} - \mu|}{s_0}\sqrt{n} \leq \frac{E}{s_0}\sqrt{n})$$
$$\geq P(\frac{|\bar{X} - \mu|}{s_0}\sqrt{n} \leq \frac{E}{s_0}\sqrt{(s_0\frac{t_{n_0-1,1-\alpha}}{E})^2})$$
$$= P(\frac{|\bar{X} - \mu|}{s_0}\sqrt{n} \leq t_{n_0-1,1-\alpha})$$
$$= 1 - \alpha$$

We use the following examples to illustrate the use of the above theorem.

Example 11.4 *Apply the two-stage confidence procedure for the following questions.*

1. Use the following data to estimate the mean stock trading price with 95% confidence level. If we need to have the accuracy at 0.05, how many observations do we need? 14.98, 15.09, 15.12, 15.15, 15.22, 14.98, 14.96

2. If we use the previous data to be the initial data and want to have a confidence estimate with accuracy at 0.06, how many additional observations do we need?

Solution. Figure 11.6 contains R codes and outputs for example 11.4. In the seven-day stock exchange data, the sample mean is \$15.07 with sample standard deviation of \$0.10. At 95% confidence level, the mean stock exchange price is estimated in the range from \$14.98 to \$15.16. The estimated error is \$0.09, which is beyond the 6 cents requirement. Using the two-stage confidence approach as in Theorem 11.2, a total of 24 observations (additional 17 observations) is needed to keep the confidence level at 0.95 and accuracy level at 6 cents.

11.2   Simultaneous Learning in Dose-response Studies

When data come from two or more resources, the random error in each resource attributes to the prediction error in the overall statement combining the multiple resources. In this case, it is desirable to control the error rates simultaneously so that the overall error rate can be maintained at a nominated level. In statistics and data analysis, simultaneous inference contains methodologies controlling errors from various resources. In the earlier time, this includes the conventional methods such as all pairwise multiple comparisons by John Tukey[118], multiple comparisons with a control by Dunnett[45], and multiple

```
stockdata <- c(14.98, 15.09, 15.12, 15.15, 15.22, 14.98, 14.96)

TwoStageConfidenceSize <- function(x, alpha,d){
  n0 <- length(x)
  xbar0 <- mean(x)
  s0 <- sd(x)
  t0 <- qt(1-(alpha/2), n0-1)
  CI <- xbar0+c(-1,1)*(s0/sqrt(n0))*t0
  error <- (CI[2]-CI[1])/2
  z <- (t0/d)**(-2)
  n <- (s0**2)/z
  n <- ceiling(n)
  return(list(SampleMean=xbar0,SampleSD=s0, t0=t0, CI=CI,error=error,z=z,n=n))
}

TwoStageConfidenceSize(stockdata,0.05,0.05)

$SampleMean
[1] 15.07143

$SampleSD
[1] 0.1000714

$t0
[1] 2.446912

$CI
[1] 14.97888 15.16398

$error
[1] 0.09255061

$z
[1] 0.0004175451

$n
[1] 24
```

FIGURE 11.6

Sample size calculation for stock data-3

comparisons with the best by Hsu [67], as well as neighboring comparisons by Chen and Hoppe[28], to list just a few. More details in this regard can be found in the book by Hochberg and Tamhane (1987) as well as Hsu (1996), methods associated with probability inequalities can also be found, for example, in Chen (2014) [22]. In what follows in this section, we shall discuss some current developments on partitioning methods for multiple comparisons, which is related to decision trees and range regression that we discussed in Chapter 9. Under the condition of directed and inverted confidence regions, the first subsection is on step-up simultaneous confidence procedures to identify the minimum effective dose of a drug. The second subsection is on simultaneous confidence band estimating several dose-response curves spontaneously over a continuous domain of the input dosage.

11.2.1 Confidence Procedures for Aspirin Efficacy

We start with an illustrative example to introduce the setting and the information background on step-up confidence procedures.

Example 11.5 *Consider a double-blinded experiment for the efficacy of Aspirin regarding acute myocardial infarction in men with unstable angina. 60 patients were randomly assigned to five distinctive dosages (for instance, 100mg, 200mg, 250mg, 300mg, and 325mg, daily) for a period of time, where the 100mg Aspirin treatment serves as the baseline (active control) group. Patients in the treatment groups of 250mg and 325mg were given Heparin (5000-U intravenous bolus) as a supplement treatment in conjunction with Aspirin. After diagnostic evaluations, each patient was assigned a risk score based on the updated health information including the diagnosis of pathologic Q-wave changes on electrocardiograms.*

The outcome variable is the cardiovascular safety score ranging from 1 to 20, indicating the highest to the lowest risk of heart attach in the future. the goal of the study was to detect treatments that effectively reduce future heart-attack risks by significantly increasing cardiovascular safety scores.

In the study, a treatment is regarded as being significant if it improves the median cardiovascular risk score by 5 units, compared with that of the active control group.

Without a specific model assumption for the underlying distribution of the data, we use Wilcoxon's rank sum statistic to test the median score difference, $H_i : \eta_i - \eta_0 \leq 5$ versus $K_i : \eta_i - \eta_0 > 5$ for $i = 1, 2, 3, 4$, of the risk score difference between the treatment group and the baseline group ($i = 0$ represents the treatment of 100mg daily, the active control group). From the data set, the pairwise Wilcoxon test statistics between the treatment and the baseline group are, $W_1 = 75.5$, $W_2 = 78$, $W_3 = 76$, $W_4 = 82$, and the p-values of pairwise comparisons are $\hat{p}_1 = 0.0142$, $\hat{p}_2 = 0.0226$, $\hat{p}_3 = 0.0156$, and $\hat{p}_4 = 0.0445$. Thus, the ordered p-values corresponding to the changes of cardiovascular scores are $\hat{p}_{(1)} = 0.0142$, $\hat{p}_{(2)} = 0.0156$, $\hat{p}_{(3)} = 0.0226$, and $\hat{p}_{(4)} = 0.0445$.

The above methods of analysis fail to identify the efficacy of Aspirin combined with Heparin, which is clinically regarded as an efficacious treatment. Neither the Bonferroni adjustment nor the Holm's procedure is able to detect the significant difference, given the fact that the smallest p-value (0.0142) is larger than the cut-off value, $\alpha/4 = 0.0125$. On the other hand, the application of Hochberg's step-up procedure requires additional model assumptions (such as MTP_2), which are implausible to verify for this set of data. Under this scenario, a new inference methodology is called.

Before discussing confidence procedures, we shall introduce two basic concepts in the sequel. The first one is the concept of *inverted confidence set*, and the second one is the concept of *directed confidence set*.

Definition 11.3 Inverted Confidence Set. *For $\boldsymbol{\theta} = (\theta_1, ..., \theta_k)^T$, let $\hat{P}_i(\mathbf{y}|\theta_i^*)$ be the p-value for the simple null hypothesis $H_{i0}^* : \theta_i = \theta_i^*$ versus $\theta_i \in \Theta_i^c$, where $\theta_i^* \in \Theta_i$. Denote $C_i^t(\mathbf{y})$ (or C_i^t for notational convenience) the set $\{\mathbf{y} : \boldsymbol{\theta} \in \{\hat{P}_i(\mathbf{y}|\boldsymbol{\theta}) \geq \alpha/t\}\}$ for any integer $t = 1, ..., k$. (Here, we use the notation in Casella and Berger, 2002 [16], page 463). Since this confidence set* *is actually inverted from the corresponding rejection region via the individual p-value, we name it inverted confidence set.*

The following example illustrates the concept of an inverted confidence set $C_i^t(\mathbf{y})$ when the data comes from a set of k normal populations.

Example 11.6 *Consider a set of data from k different resources, $\mathbf{y} = (X_{ij})$, $i = 1, ..., k$ and $j = 1, ..., n_i$, where $X_{ij} \sim N(\mu_i, \sigma_i^2)$ and σ_i is unknown. For a fixed index i, consider the null hypothesis $H_{i0} : \mu_i \leq \mu_{i0}$ versus the alternative hypothesis $H_{i1} : \mu_i > \mu_{i0}$. Denote T_v the corresponding t-statistic, and $t_{1-\alpha}$ the value satisfying $P(T_v < t_{1-\alpha}) = 1 - \alpha$. Also denote corresponding rejection region*

$$R = \{\mathbf{y} : (\overline{X}_i - \mu_{i0})/(\hat{\sigma}_i/\sqrt{n_i}) > t_{1-\alpha}\}.$$

We are interested in finding the inverted confidence set with confidence level $1 - \alpha$.

For any value $\mu_i^* \leq \mu_{i0}$, the p-value for $\mu_i = \mu_i^*$ versus $H_{i1} : \mu_i > \mu_{i0}$ is

$$\hat{P}_i(\mathbf{y}|\mu_i^*) = P(T_v > (\overline{X}_i - \mu_i^*)/(\hat{\sigma}_i/\sqrt{n_i}));$$

and the p-value for $H_{i0}^* : \mu_i \leq \mu_{i0}$ versus $H_{i1} : \mu_i > \mu_{i0}$ is

$$\begin{aligned}
\hat{P}_i(\mathbf{y}) &= \hat{P}_i(\mathbf{y}|\Theta_{i0}) \\
&= \sup_{\theta_i \in \Theta_{i0}} \hat{P}_i(\mathbf{y}|\theta_i) \\
&= P(T_v > (\overline{X}_i - \mu_{i0})/(\hat{\sigma}_i/\sqrt{n_i})).
\end{aligned}$$

Now, notice that for any value μ_i, $\hat{P}_i(\mathbf{y}|\mu_i)$ has the following property:

$$\begin{aligned}
&\{\mathbf{y} : \mu_i \in \{\hat{P}_i(\mathbf{y}|\mu_i) \geq \alpha\}\} \\
&= \{\mathbf{y} : \mu_i \in \{(\overline{X}_i - \mu_i)\sqrt{n_i}/\hat{\sigma}_i \leq t_{1-\alpha}\}\} \\
&= \{\mathbf{y} : \mu_i \in \{\mu_i \geq \overline{X}_i - t_{1-\alpha}(\hat{\sigma}_i/\sqrt{n_i})\}\}.
\end{aligned}$$

Thus, the inverted confidence set for μ_i is

$$C(\mathbf{y}) = (\overline{X}_i - t_{1-\alpha}(\hat{\sigma}_i/\sqrt{n_i}), \quad +\infty),$$

and for each μ_i,

$$\begin{aligned}
&P(\mathbf{y} : \hat{P}_i(\mathbf{y}|\mu_i) \geq \alpha) \\
&= P((\overline{X}_i - \mu_i)\sqrt{n_i}/\hat{\sigma}_i \leq t_{1-\alpha}) \\
&= 1 - \alpha.
\end{aligned}$$

Another concept we need in this section is the *Directed Confidence Set* (Hsu and Berger, 1999 [68]): A confidence set for θ, $C(\mathbf{y})$, is said to be directed toward a subset Θ^* of the parameter space Θ, if for every sample point $\mathbf{y}$, either $\Theta^* \subset C(\mathbf{y})$ or $C(\mathbf{y}) \subset \Theta^*$.

Example 11.7 *In Example11.6, the confidence interval for*

$$C(\mathbf{y}) = \{\mathbf{y} : \mu_i \in (\overline{X}_i - t_{1-\alpha}(\hat{\sigma}_i/\sqrt{n_i}), \quad +\infty)\}$$

is directed toward the alternative parameter space

$$\Theta^* = \{\mu : \mu_i > \mu_{i0}\}.$$

This is because either the confidence set $C(\mathbf{y})$ contains the alternative space Θ^*, when

$$\mu_{i0} > \overline{X}_i - t_{1-\alpha}(\hat{\sigma}_i/\sqrt{n_i}),$$

or it is contained in the alternative space, when

$$\mu_{i0} < \overline{X}_i - t_{1-\alpha}(\hat{\sigma}_i/\sqrt{n_i}).$$

For a given sample, the confidence set is a subset in the parameter space. For a given parameter, the confidence set is an event in the sample space (see, for example, Berger and Casella, 2002, p 463[16]). The concept of directed confidence set, in conjunction with the inverted confidence set, leads to the following result.

For multiple testing problem of $H_{i0} : \theta_i \in \Theta_i$ versus $H_{i1} : \theta_i \in \Theta_i^c$, assume that for any nested rejection region and any permissible integers i and t, there exists an inverted confidence set $C_i^t(\mathbf{y})$ that is directed towards Θ_i^c. When screening down from the largest to the smallest ordered p-value, let m be the index that satisfies the following two criteria,

i) $\hat{P}_{(m)} \geq \alpha/(k - m + 1)$; and

ii) for any index i: $m < i \leq k$, $\hat{P}_{(i)} < \alpha/(k - i + 1)$. For notational convenience, denote

$$C_0^k(\mathbf{y}) = \Theta$$

when $m = 0$ where all p-values are smaller than the corresponding cutoff values, and

$$\Theta_{(k+1)}^c = \Theta$$

when $m = k$. Under this setting, the simultaneous confidence set keeps the confidence level at the nominal level.

Theorem 11.3

$$P(\boldsymbol{\theta} \in \Theta_{(k+1)}^c \bigcap \Theta_{(k)}^c \bigcap \cdots \bigcap \Theta_{(m+1)}^c \bigcap C_{(m)}^{k-m+1}(\mathbf{y})) \geq 1 - \alpha. \qquad (11.1)$$

The proof of the theorem can be found in Chen(2016 [23]). Denote

$$\alpha(i) = \alpha/(k - i + 1),$$

Theorem 11.3 leads to a step-up operating algorithm for the construction of simultaneous confidence sets. We will describe the new confidence procedure and apply the procedure to construct simultaneous confidence sets for the Aspirin example.

Algorithm: A Step-up Confidence Procedure

Step 1	If $\hat{P}_{(k)} > \alpha$,	then claim $\theta_{(k)} \in \{\hat{P}_{(k)} \geq \alpha\}$, stop; else, claim $\theta_{(k)} \in \Theta^c_{(k)}$, and go to Step 2.
Step 2	If $\hat{P}_{(k-1)} > \alpha/2$,	then claim $\theta_{(k-1)} \in \{\hat{P}_{(k-1)} \geq \alpha/2\}$, stop; else claim $\theta_{(k-1)} \in \Theta^c_{(k-1)}$, go to Step 3.
	$\vdots$	
Step i	If $\hat{P}_{(i)} > \alpha(i)$,	then claim $\theta_{(i)} \in \{\hat{P}_{(i)} \geq \alpha(i)\}$, stop; else claim $\theta_{(i)} \in \Theta^c_{(i)}$, and go to Step $i+1$.
	$\vdots$	
Step k	If $\hat{P}_{(1)} > \alpha/k$,	then claim $\theta_{(1)} \in \{\hat{P}_{(1)} \geq \alpha/k\}$, stop; else claim $\theta_{(1)} \in \Theta^c_{(1)}$, stop.

We now apply the algorithm developed above to analyze the Aspirin efficacy example in Example 11.5. Notice that the alternative parameter space is of the form $\theta_i = \eta_i - \eta_0 \leq 5$, and the associated confidence region corresponding to $\hat{p}_i \geq \alpha/t$ is of the form $W \leq c_{1-\alpha}$, where W is the Wilcoxon statistic. Thus the inverted confidence set is of the form $\eta_i - \eta_0 \geq U$, where U is a lower confidence bound for the median difference derived from W, and the condition of directed confidence interval is satisfied.

By the step-up confidence set algorithm, we have the following analytical results when $\alpha = 0.05$.

Algorithm for the Aspirin Example

Step 1	$\hat{P}_{(4)} = 0.0445 < \alpha$, claim $\theta_{(4)} \in \Theta^c_{(4)}$, and go to Step 2.
Step 2	$\hat{P}_{(3)} = 0.0226 < \alpha/2$, claim $\theta_{(3)} \in \Theta^c_{(3)}$ and go to Step 3.
Step 3	$\hat{P}_{(2)} = 0.0156 < \alpha/3$, claim $\theta_{(2)} \in \Theta^c_{(2)}$ and go to Step 4.
Step 4	$\hat{P}_{(1)} = 0.0142 > \alpha/4$, claim $\mu_{(1)} - \mu_0 \in \{\hat{P}_{(1)} \geq \alpha/4\}$, stop.

Thus, the 95% simultaneous confidence set consists of the following components:

$$\eta_4 - \eta_0 - 5 \in (0, \infty)$$
$$\eta_3 - \eta_0 - 5 \in (0, \infty)$$
$$\eta_2 - \eta_0 - 5 \in (0, \infty)$$
$$\eta_1 - \eta_0 - 5 \in (-0.002, \infty).$$

The above prediction outcomes can be interpreted as follows. Different from the previous inference procedures, the step-up confidence procedure is able to claim that the Aspirin treatments significantly reduce the risk of heart attack in the clinical trial. The improvement of the cardiovascular safety score is at least five units for the first three treatments, with the following statements simultaneously.

1. The daily treatment of Aspirin at 250 mg is statistically significant in reducing the risk of heart attack. It increases the cardiovascular safety score by at least 5 units.

2. The daily treatment of Aspirin at 300mg in conjunction with supplement treatment of Heparin is statistically significant in reducing the risk of heart attack. It increases the cardiovascular safety score by at least 5 units.

3. The daily treatment of Aspirin at 325 mg with Heparin is statistically significant in reducing the risk of heart attack. It increases the cardiovascular safety score by at least 5 units.

4. The daily treatment of Aspirin at 200 mg along improves the median cardiovascular score by 4.998 units. This is because the $1 - \alpha/4 = 98.75\%$ confidence interval for the median difference is

$$\eta_1 - \eta_0 - 5 \in (-0.002, \infty).$$

The learning result of the new step-up confidence method fits well with clinical expectations. Chen (2016)[23] contains more technical discussions on motivation, theory, and applications of this method.

11.2.2　Confidence Bands on Thrombolysis Effects

Since a dose-response curve is usually continuous, it is desirable to have a simultaneous confidence band instead of confidence intervals in dose-response studies. In this section, we describe a method proposed by Kerns and Chen (2017) [76] on the construction of simultaneous confidence bands. We present a method to construct simultaneous confidence bands for the comparison of multiple regression lines when individual confidence bands are available. The theoretical result will be illustrated by an example comparing efficacy curves of successful rates of thrombolysis for three different patient groups. Technical details including theoretical proofs and simulations can be found in the paper by Kerns and Chen (2017) [76].

In this subsection, we shall describe a general theorem on the construction of step-wise simultaneous confidence bands, delineate the operating algorithm, and illustrate the method with an example on impacts of theormbolysis for different age groups of patients.

In a logistic regression model

$$y(x_1, ..., x_m) = P(Y = 1 | X_i = x_i, i = 1, \ldots, m)$$
$$= \frac{1}{1 + exp[-(\beta_0 + \beta_1 x_1 + \cdots + \beta_m x_m)]},$$

where x_i $i = 1, ..., m$ are continuous variables such as time of exposure to a risk factor or the dosage of a drug.

Consider a special case where $m = 1$ for the input of drug dosage with k groups of treatments. Let $\hat{y}_i(x)$ denote a $100(1 - \alpha)\%$ lower confidence bound for the response curve $y_i(x), i = 1, 2, \ldots, k$, and let $\delta(x)$ denote a threshold on the efficacy of the drug.

$$\hat{y}_i^*(x) = \begin{cases} \min(\hat{y}_i(x), \quad \delta(x)), \\ \qquad \text{if there exists a value } x_0 \text{ such that } \hat{y}_i(x_0) < \delta(x_0) \\ \hat{y}_i(x), \quad \text{if } \hat{y}_i(x) > \delta(x) \text{ for all } x. \end{cases}$$

(11.2)

Denote the set $L_i(x) = (\hat{y}_i^*(x), \infty)$. The set $L_i(x)$ possesses the following statistical properties.

First, the one-sided confidence region is directed toward the pre-specified efficacy region $(\delta(x), \infty)$. For any $i = 1, 2, \ldots, k$, if the boundary $\hat{y}_i(x) < \delta(x)$ at a point $x = x_0$, then $\hat{y}_i^*(x) < \delta(x)$. Therefore, the set L_i is directed toward the set $\Delta(x) = (\delta(x), \infty)$, that is, either $L_i(x) \subseteq \Delta(x)$ or $\Delta(x) \subseteq L_i(x)$.

Second, the one-sided confidence region $L_i(x)$ reaches the nominal confidence coverage. For any $i = 1, 2, \ldots, k$, if the boundary $\hat{y}_i(x)$ is a $100(1 - \alpha)\%$ lower confidence bound for the response curve $y_i(x)$, then $\hat{y}_i^*(x)$ is also a $100(1 - \alpha)\%$ lower confidence bound for $y_i(x)$.

We need the following notations for the key theorem in this subsection.

1. Simultaneous confidence lower bands. Consider k population of interest. For any $i = 1, 2, \ldots, k$, denote

$$L_i(x) = (\hat{y}_i^*(x), \infty)$$

a $100(1 - \alpha)\%$ lower confidence set for the logistic regression line $y_i(x)$. Assume that $L_i(x)$ is directed toward the set of alternative space $\Delta(x)$. Also, let

$$\hat{y}_{k+1}^*(x) = \min_i \hat{y}_i^*(x),$$

which is a $100(1 - \alpha)\%$ lower confidence bound for the lower boundary across all the response curves, $\min_i y_i(x)$.

Let D be the smallest integer i such that $\hat{y}_i^*(x) \not> \delta(x)$ if such $i(1 \leq i \leq k)$ exists; Otherwise, let $D = k + 1$.

2. Simultaneous confidence upper bands Suppose that

$$U_i(x) = (-\infty, \hat{y}_i^{**}(x)), i = 1, 2, \ldots, k,$$

is a $100(1 - \alpha)\%$ upper simultaneous confidence band for the parameter function $y_i(x)$ that is directed toward the set of parameter

$$\Lambda_i^c(x) = (-\infty, \eta(x)),$$

respectively. Let

$$\hat{y}_{k+1}^U(x) = \max_i \hat{y}_i(x).$$

For the parameter array of logistic regression functions, $\theta(x)$.

Let T denote the smallest integer i such that $\hat{y}_i^{**}(x) \not< \eta(x)$ if such $i (1 \le i \le k)$ exists; Otherwise, let $T = k + 1$.

3. Simultaneous two-sided confidence bands Let $M = \min(T, S)$, and define the set

$$Y^W(x) = \begin{cases} (\Theta_1^c \cap \Lambda_1^c) \cap (\Theta_2^c \cap \Lambda_2^c) \cap \ldots \cap \ (\Theta_{M-1}^c \cap \Lambda_{M-1}^c) \cap L_S(x), \\ \qquad\qquad \text{if } S < T, \\ (\Theta_1^c \cap \eta_1^c) \cap (\Theta_2^c \cap \eta_2^c) \cap \ldots \cap \ (\Theta_{M-1}^c \cap \eta_{M-1}^c) \cap U_T(x), \\ \qquad\qquad \text{if } S > T \end{cases}$$

(11.3)

With the above setting, we have the following result.

Theorem 11.4 *For all* $\theta(x) \in \Theta$, *we have simultaneous confidence lower bands,*

$$P_{\theta(x)}(\theta(x) \in \Theta_1^c \cap \Theta_2^c \cap \ldots \cap \Theta_{S-1}^c \cap L_D(x), \quad x \in (a, b)) \ge 1 - \alpha.$$

Simultaneous confidence upper bands,

$$P_{\theta(x)}(\theta(x) \in \Lambda_1^c \cap \Lambda_2^c \cap \ldots \cap \Lambda_{T-1}^c \cap U_T(x), x \in (a, b)) \ge 1 - \alpha.$$

and simultaneous confidence two-sided bands, for all $\theta(x) \in \Theta$,

$$P_{\theta(x)}(\theta(x) \in Y^W(x), x \in (a, b)) \ge 1 - \alpha,$$

where $Y^W(x)$ *is a set of two-sided* $100(1 - \alpha)\%$ *confidence bounds for* $\theta(x)$.

The proof of the above result can be found in Kerns and Chen (2017)[76]. Notice that both $L_S(x) \cap \Lambda_S^c$ when $S < T$ and $U_T(x) \cap \Theta_T^c$ when $S > T$ are $100(1 - \alpha)\%$ confidence bounds for $\theta(x)$.

In order to use the above theorem to construct simultaneous confidence bands, in the first step we need to establish an individual confidence band for each single logistic regression curve under consideration. Methods in the construction of individual confidence band can be found in Piegorsch and Casella (1988) [93]), among others. After obtaining individual confidence bands, in the second step, we can form simultaneous confidence bands $\hat{y}_i^*(x)$, for $i = 1, ..., k$ by deriving confidence bands directed toward the alternative parameter space, as described above.

For the easy of application in the construction of simultaneous confidence bands for k logistic regression lines, we can simplify the above theorem into an operating algorithm. Notice that the parameter array of logistic regression lines $\theta(x)$ is now a function based on the value x, and the directed toward confidence set is now a region bounded by a function $\theta(x)$, on the basis of the individual confidence bands, the stepwise confidence procedure corresponding to the above theorem can be formulated into the following procedure.

Stepwise Lower Confidence Bands Procedure

Step 1	If $\hat{y}_1^*(x) > \delta(x), x \in (a, b)$, then assert $y_1(x) > \delta(x), x \in (a, b)$, and go to Step 2; else assert $y_1(x) > \hat{y}_1^*(x), x \in (a, b)$ and stop.
Step 2	If $\hat{y}_2^*(x) > \delta(x), x \in (a, b)$, then assert $y_2(x) > \delta(x), x \in (a, b)$, and go to Step 3; else assert $y_2(x) > \hat{y}_2^*(x), x \in (a, b)$ and stop.
$\vdots$	
Step k	If $\hat{y}_k^*(x) > \delta(x), x \in (a, b)$, then assert $y_k(x) > \delta(x), x \in (a, b)$, and go to Step k + 1; else assert $y_k(x) > \hat{y}_k^*(x), x \in (a, b)$ and stop.
Step k + 1	Assert $\theta(x) \in \hat{y}_{k+1}^*(x), x \in (a, b)$ and stop.

Upon the establishment of the above theorem and algorithm, we may now illustrate their applications using the following example.

TABLE 11.1

Success rates of thrombolytic therapy

Lysis time	0.25	1.25	2.5	8	24	50
Log-dose	-1.386	0.223	0.916	2.08	3.18	3.91
Group 1	16/78	23/78	48/78	56/78	68/78	78/78
Group 2	10/78	13/78	37/78	48/78	54/78	65/78
Group 3	2/78	11/78	22/78	25/78	35/78	44/78

Example 11.8 *In a prospective randomized trial designed to investigate the comparative results of traditional surgical revascularization with those of catheter-directed thrombolytic therapy, the therapy is intra-arterial catheter directed thrombolysis with UK (Urokinase) 250,000 unit bolus followed by 4000 units/min for 4 hours, and then 2000 units/min for as many as 36 hours.*

To evaluate the age impact on the successful rate of thrombolysis response curves, three categories of patients were studied. The first group consists of patients whose ages were under 55 and had acute limb ischaemia (onset of symptoms < 14 days); the second group consists of patients whose ages were under 55 and have chronic limb ischaemia (onset of symptoms > 14 days);

and the third group consists of patients whose ages were over 55 with (acute or chronic) limb ischaemia. The dose variable in this example is the time of therapy and the outcome is the success rate of the procedure (more than 80% lysis). For illustration purpose, summaries of data information are given in Table 11.1.

Based on the information in Table 11.1, the ML estimates from the logistic fit and the Fisher information matrix for each group are found and displayed in the following output.

TABLE 11.2
Parameter estimation on dose-response curves

Group 1	$\beta = [\ -.8699 \quad .2654\ ]$	$F = \begin{bmatrix} 162.12 & 328.68 \\ 328.68 & 1103.38 \end{bmatrix}$
Group 2	$\beta = [\ -1.1374 \quad .3179\ ]$	$F = \begin{bmatrix} 132.81 & 284.09 \\ 284.09 & 944.85 \end{bmatrix}$
Group 3	$\beta = [\ -1.3244 \quad .3363\ ]$	$F = \begin{bmatrix} 84.33 & 193.69 \\ 193.69 & 646.02 \end{bmatrix}$

Table 11.2 was used to construct the individual confidence bands as in Kerns and Chen (2017)[76].

Figure 11.7 displays the estimated 95% lower confidence bands for three groups, along with the threshold, which is chosen to be $\delta(x) = 1/\{1 + exp[-(-1.39 + .3x)]\}$. Three lower confidence bands are displayed using dotted lines, with the order of Group 1, Group 2, Group 3, from top to bottom. The thick solid line is for the threshold. It is indicated by the figure that the therapy is effective for patients in the first age group only.

In this example, we consider the analysis of success rates of thrombolysis for patients with DVT (deep vain thrombosis). Thrombosis occurs when blood cells and proteins (e.g., erythrocytes, leukocytes, platelets, and fibrin) aggregate and clot within an intact vein. Thrombolytic therapy is a treatment procedure offering a compromise between surgery and anticoagulation. It provides a potential to restore arterial flow through mildly invasive techniques. Despite the theoretical advantages of the thrombolysis procedure, the safety and efficacy of thrombolysis are often debated in the medical literature. One of the safety concerns is that thrombolysis may restore blood flow more slowly than immediate surgical revascularization. In that case, tissue ischemia may progress to infarction before the artery has recanalized, hence result in internal bleeding, stroke, and other complications. More information on the efficacy of thrombolysis therapy can be found in Weaver et al. (1996, [124]), Aziz, Chen and Comerota (2011, [4]), and Chen and Comerota (2012, [24]).

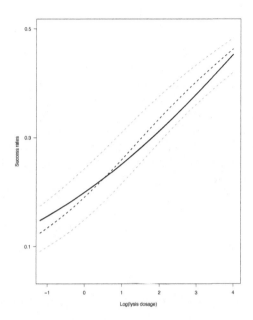

FIGURE 11.7
Simultaneous lower bands for thrombolysis effects

11.3 Weighted Simultaneous Confidence Regions

Besides simultaneous confidence sets at different dosage level and simultaneous confidence bands for two or more continuous logistic regression curves, another critical aspect in simultaneous learning is weighting inference. As pointed out in Ma et al. (2023 [82]), when two or more confidence regions are involved in a study, it is necessary to have the method of weighted simultaneous confidence regions. For example, in deep learning where weight information is cooperated into different layers in the prediction process. Population weight is an integrated part of multiple inference. For simplicity and convenience in discussions, we traditionally assume equal weights for populations under investigation, which is not necessarily true all the time. For instance, in a system consisting of multiple subsystems, when some systems are relatively newer or more efficient than others, the difference on ages of the subsystems may result in different levels of reliability performance. It is unrealistic to assume that all the subsystems carry the same weight in the evaluation of the system reliability. Similar applications also occur in various scenarios. For instance, in gene micro-array analysis when some genes are biologically more closely (than others) related to a specific disease, we need to assign higher weights when performing simultaneous inference on those genes; in system reliability analysis in engineering, where the reliability of primary subsystems is tested prior to the regular (or secondary) subsystems, we need to put relatively higher weights on the primary subsystems. In investment strategy analysis in actuarial science, when certain stocks have dominating priorities in building investment portfolios, we increase weights for prioritized stocks in portfolio analytics. The conventional way of assigning equal weights to all involved populations is not always valid in practice.

In the rest of this section, we shall synthesize two studies for recent methodological developments on simultaneous inference for weighted hypotheses. The first one is on the analysis of gene expression data for breast cancer investigation. The second subsection is on the inference for parameter arrays, a new concept expanding the inference on vector-style multivariate parameters to matrix-style parameter arrays. By increasing more parameter dimensions, and each dimension contains infinitely many values, the method by Ma et al. (2023)[82], essentially opens a new direction in high dimensional simultaneous inference.

11.3.1   Weighted Hypotheses and Breast Cancer Study

We start with an application of weighted hypothesis in breast cancer study. Instead of blindly assigning equal weights to each gene under investigation, we use additional biological information to assign different weights according to their relevance to the disease, when selecting genes with significantly different expressions. The new procedure also strongly controls the family-wise error rate in multiple-path learning.

Storey and Tibshirani (2003, [114]) used the false discovery rate (FDR) to identify genes associated with breast cancer. The breast cancer susceptibility genes, BRCA1 and BRCA2, play a critical role in DNA repair and apoptosis. If either one mutates, its function of inhibiting tumor genesis would be affected and leads to an elevated risk of breast cancer. Together, BRCA 1 and BRCA2 mutations account for a lifetime risk of 50 to 85 percent of breast cancer and 15 to 45 percent of ovarian cancer (Hedenfalk et al. 2001) [57]. Hedenfalk et al. (2001) examined tissue samples from BRCA1-mutation-positive, BRCA2-mutation-positive tumors, and some sporadic cases to detect differential gene-expressions among these three types of breast cancer.

Based on a modified F-test, 51 of the 3,226 genes were determined to be significant with a threshold $\alpha = .001$. Storey and Tibshirani (2003) conducted related research using a subset of genes ($k = 3,170$). With a cutoff $q < .05$ for the control of false discovery rate (FDR), they found 160 significant genes for differential expression between BRCA1- and BRCA2- mutation-positive tumors. The data set is available as the supplementary material in Storey and Tibshirani (2003, [114]).

Example 11.9 *The inference question in this example is the following. When* *additional biological information on breast cancer is involved in the analysis for the dataset in Storey and Tibshirani(2003), how to obtain the prediction result with strong control of family-wise error rate, instead of FDR?*

The problem can be formulated as follows. Let $\boldsymbol{\theta} = (\theta_1, \theta_2, ..., \theta_k)^T$ be a vector of parameters and H_1, H_2, ..., H_k be k hypotheses with null hypothesis $H_{i0} : \theta_i \in \Theta_i$ versus $H_{i1} : \theta_i \in \Theta_i^c$. Assume that prior information is available such that the hypotheses can be weighted according to the order of importance. Denote by c_i the weight for H_i, $i = 1, 2, ..., k$, such that larger c_i's correspond to more important hypotheses. Let

$$w_t = \sum_{i \in I_t} c_i,$$

where I_t is an index set for hypotheses to be tested simultaneously.

Given a set of data $\mathbf{y}$, let $\hat{P}_1$, $\hat{P}_2$, ..., $\hat{P}_k$ denote the p-values corresponding to tests ϕ_1, ϕ_2, ..., ϕ_k for H_1, H_2, ..., H_k, respectively. Different from the weighted step-down testing procedure of Holm (1979) [60] and weighted step-up procedure of Tamhane and Liu (2008)[115] , we propose a weighted step-down confidence set procedure in this subsection.

For convenience, consider a null hypothesis

$$H_{i0}^* : \theta_i \geq \theta_i^* \quad versus \quad H_{i1}^* : \theta_i < \theta_i^*,$$

and denote the corresponding p-value by $\hat{P}_i(\mathbf{y}|\theta_i^*)$. Let $C_i^t(\mathbf{y})$ be an inverted confidence set

$$C_i^t(\mathbf{y}) = \{\mathbf{y} : \boldsymbol{\theta} \in \{\hat{P}_i(\mathbf{y}|\boldsymbol{\theta}) \geq c_t \alpha / w_t\}\}$$

for any $t = 1, 2, ..., k$.

For a multiple testing problem of $H_{i0} : \theta_i \in \Theta_i$ versus $H_{i1} : \theta_i \in \Theta_i^c, i = 1, 2, ..., k$, assume that there exists an inverted confidence set $C_i^t(\mathbf{y})$ that is directed towards Θ_i^c for all permissible integers i and t, then for the index m in the step-down weighted procedure, denote $\Theta_{(0)}^c = \Theta$ for notation convenience, $\Theta_{(i)}^c$ is the alternative space associated with $S_{(i)}$, and $C_{(m)}^{k-m+1}(\mathbf{y})$ is the inverted and directed confidence set that is associated with $\Theta_{(m)}^c$.

Theorem 11.5 *With above settings, we have*

$$P\left(\mathbf{y} : \boldsymbol{\theta} \in \Theta_{(0)}^c \bigcap \Theta_{(1)}^c \bigcap \cdots \bigcap \Theta_{(m-1)}^c \bigcap C_{(m)}^{k-m+1}(\mathbf{y})\right) \geq 1 - \alpha, \quad (11.4)$$

Theorem 11.5 is a step-wise confidence procedure taking into consideration of weights for the importance of hypotheses. It shows that once we can find individual confidence sets adjusted by the Bonferroni inequality in conjunction with the property of directed toward the corresponding alternative space, we can derive the simultaneous confidence sets with the nominal overall confidence level. The theorem can be formulated into a prediction algorithm as follows.

Algorithm: Weighted Step-down Confidence Set Procedure

Step 1: If $S_{(1)} \geq \frac{\alpha}{w_k}$, then conclude $\theta_{(1)} \in \{\theta : \hat{P}_1^* \geq \frac{c_1^* \alpha}{w_k}\}$, stop; else conclude $\theta_{(1)} \in \Theta_{(1)}^c$, and go to Step 2.

Step 2: If $S_{(2)} \geq \frac{\alpha}{w_{k-1}}$, then conclude $\theta_{(2)} \in \{\theta : \hat{P}_2^* \geq \frac{c_2^* \alpha}{w_{k-1}}\}$, stop; else conclude $\theta_{(2)} \in \Theta_{(2)}^c$, go to Step 3.

$\vdots$

Step i: If $S_{(i)} \geq \frac{\alpha}{w_{k-i+1}}$, then conclude $\theta_{(i)} \in \{\theta : \hat{P}_i^* \geq \frac{c_i^* \alpha}{w_{k-i+1}}\}$, stop; else conclude $\theta_{(i)} \in \Theta_{(i)}^c$, and go to Step $i+1$.

$\vdots$

Step k: If $S_{(k)} \geq \frac{\alpha}{w_1}$, then conclude $\theta_{(k)} \in \{\theta : \hat{P}_k^* \geq \alpha\}$, stop; else conclude $\theta_{(k)} \in \Theta_{(k)}^c$, stop.

We shall now use an example to illustrate the learning algorithm discussed above. First, we shall discuss an example to explain the concept of *Inverted Confidence Set* $C_i^t(\mathbf{y})$ for the difference of two normal population means. More details on the methodology can be found in Yu et al. (2022, [128]) or Casella and Berger (2002)[16].

Example 11.10 *We consider a data set (X_{ijt}), $i = 1, 2$ for treatment versus placebo, $j = 1, 2, ..., n_i$ for experimental subjects, and $t = 1, 2, ..., k$ for k different treatments, where X_{ijt}'s follow independent $N(\mu_{it}, \sigma_t^2)$ with an unknown standard deviation σ_t. We are interested in testing whether there is a treatment effect of t unit difference between the treatment and placebo for each Treatment-t,*

$$H_{0t} : \mu_{1t} \geq \mu_{2t} + t, \quad versus \quad H_{1t} : \mu_{1t} < \mu_{2t} + t,$$

for $t = 1, .., .., k$. Depending on the importance of each hypothesis, we assign weights $c_1, c_2, ..., c_k$ successively and let

$$S_t = \frac{\hat{P}_t}{c_t}$$

for $t = 1, 2, ..., k$.

Let $S_{(1)} \le S_{(2)} \le ... \le S_{(k)}$ be the ordered S-values and

$$S_{(t)} = \frac{\hat{P}_t^*}{c_t^*},$$

where $\hat{P}_t^*$ and c_t^* are the corresponding p-value and hypothesis weight. Comparing the ordered S-value with the corresponding significant level

$$\frac{\alpha}{w_{k-t+1}},$$

where

$$w_{k-t+1} = \sum_{s=t}^{k} c_s^*.$$

Now, we can consider the proposed weighted step-down confidence set procedure with the assumption that the first non-rejected hypothesis occurs at step q $(q \le k)$.

Based on the above theorem, the inverted confidence interval for step q is

$$C_{(q)}^{k-q+1}(\mathbf{y}) = \left\{ \mathbf{y} : d_q \in \left(-\infty, \quad \overline{X}_{1(q)} - \overline{X}_{2(q)} + t_{v,u} \hat{\sigma_{(q)}} \sqrt{\frac{1}{n_{1(q)}} + \frac{1}{n_{2(q)}}} \right) \right\},$$

where $d_q = \mu_{1(q)} - \mu_{2(q)}$, and

$$u = 1 - \frac{c_q^* \alpha}{w_{k-q+1}},$$

$\overline{X}_{1(q)}$ and $\overline{X}_{2(q)}$ are the sample means corresponding to the q-th ordered S-value. The confidence interval $C_{(q)}^{k-q+1}(\mathbf{y})$ is applied at step q (if the procedure stops at step q).

More specifically, suppose that $k = 4$,

$$\hat{P}_1 = .10, \ \hat{P}_2 = .03, \ \hat{P}_3 = .01, \ \hat{P}_4 = .08,$$

and the corresponding weights are

$$c_1 = 35, \ c_2 = 10, \ c_3 = 40, \ c_4 = 15,$$

respectively. Then, the S-values are

$$S_1 = 0.00286, \ S_2 = 0.003, \ S_3 = 0.0025, \ and S_4 = 0.0053,$$

and the correspondingly ordered S-values are

$$S_{(1)} = .00025, \ S_{(2)} = .00286, \ S_{(3)} = .00300, \ and \ S_{(4)} = .00533.$$

Consequently,
$$\hat{P}_1^* = .01, \ \hat{P}_2^* = .1, \ \hat{P}_3^* = .03, \ \hat{P}_4^* = .08,$$
and the corresponding weights are
$$c_1^* = 40, \ c_2^* = 35, \ c_3^* = 10, \ and \ c_4^* = 15,$$
respectively. Under this setting,
$$w_4 = 100, \ w_3 = 60, \ w_2 = 25, \ and \ w_1 = 15.$$

Setting $\alpha = .05$, the weighted step-down confidence set procedure proceeds as follows.

Step 1: $S_{(1)} = .00025 < .0005 = \frac{\alpha}{w_4}$. Assert that $\theta_{(1)} \in \Theta_{(1)}^c$. Go to Step 2.

Step 2: $S_{(2)} = .00286 > .00083 = \frac{\alpha}{w_3}$. Assert that $\theta_{(2)} \in \{\hat{P}_2^* \geq \frac{c_2^* \alpha}{w_3}\} = C_{(2)}^3(\mathbf{y})$, where $\frac{c_2^* \alpha}{w_3} = .02905$. Stop.

Here, since the ordered $S_{(1)}$ corresponds to S_3, $\theta_{(1)} \in \Theta_{(1)}^c$ implies that
$$\mu_{13} < \mu_{23} + 3,$$
so
$$\Theta_{(1)}^c = (-\infty, 3).$$
If, from a set of data, we have that
$$\overline{X}_{11} - \overline{X}_{21} + t_{v, 1 - \frac{c_2^* \alpha}{w_3}} \hat{\sigma}_1 \sqrt{\frac{1}{n_{11}} + \frac{1}{n_{21}}} = 0.6,$$

then, the algorithm claims, at 95% confidence level, that Treatment-3 has significant treatment result compared with the placebo group, and the difference between the drug and placebo for Treatment-1 is less than 0.6 unit; no conclusion on Treatment-2 and Treatment-4 with the current weighting. This is partly because the weights on Treatment-2 and Treatment-4 (10 and 15, respectively) are much lower than the weights for Treatment-1 and Treatment-3 (35 and 40, respectively).

With the above setting, we can now use the weighted confidence set method to analysis the question posted in Example 11.9 as follows.

Example 11.11 *Applying the new confidence procedure to re-analyze the breast cancer data.*

Storey and Tibshirani (2003) calculated the p values from permutation test by

$$p_i = \sum_{b=1}^{B} \frac{\#\{j : |t_j{}^{0b}| \geq |t_i|, j = 1, ..., 3170\}}{3170B},$$

where $B = 100$ and $i = 1, 2, ..., 3170$. More discussion on permutation tests can be found in Huang et al (2006) [69] and Kaizar et al (2011) [74]. In light of Jauffret et al (2007) [73] regarding the significance of correlation between moesin and BRCA1 associated breast cancer, we assigned weights correspondingly to analyze the gene expression data. The reanalysis consists of two parts. In the first part, we use the published p-values in Storey and Tibshirani (2003, [114]) in conjunction with the new weighted confidence theorem. The first analysis identified several additional significant genes, such as APEX nuclease (clone 417124), apoptosis-related protein 15 (clone 137836), apoptosis inhibitor 1 (clone 34852) and two ERCC-related genes (clone 323390 and clone 52666), in addition to the 160 genes identified using the q values in Storey and Tibshirani (2003).

For example, screening up from the smallest S-value to the largest S-value, the weighted p-value of *APEX nuclease* is 2.78×10^{-9}, which is smaller than the corresponding significant level 3.30×10^{-9}. We thus concluded that the corresponding confidence interval of the mean difference of the two expressions is $\Theta_i^c = (0, +\infty)$. The expression of this gene is restrained in BRCA1-mutation-positive tumors, which results in decrease in function of mediating DNA repair (APEX nuclease).

Another example is *apoptosis inhibitor 1*, the weighted p-value of apoptosis inhibitor 1 is 1.70×10^{-8}. We recognize this gene as significant after comparing it with the corresponding significant level 2.54×10^{-8}, and conclude the mean difference is within $(0, +\infty)$. The expression of this gene, involved in suppressing apoptosis, is also decreased in BRCA 1- mutation- positive tumors.

Furthermore, the new procedure also found a gene associated with *apoptosis related protein 15*, the weighted p-value is 1.48×10^{-10} which is less than the corresponding significant level 1.55×10^{-10}. This indicates the significance of apoptosis-related protein 15 after multiplicity adjustments. The procedure stops at the 166th step with an inverted confidence interval calculated from the associated p-value.

Notice that the method of data analysis in the first part is based on permuted p-values. In the second portion of the reanalysis, we assume normality for the data and use the two-sample t-test to reanalyze the data. Under the normality assumption, we recalculated p values using two-sample t-test.

Specifically, the GATA-3 gene is highly correlated with estrogen receptor, which leads to a suppressed expression of this gene in BRCA 1-mutation-positive breast cancer (Eeckhoute et al. 2007, [46]). After assigning proper weights to the genes according to biological literature, a total of 163 significant genes were identified using the proposed weighted step-down confidence set procedure, which includes GATA-binding protein 3, moesin, among others. For those genes identified as significant, we concluded that the mean difference of these genes is within $(0, +\infty)$. The procedure stops at the 164th gene where the mean difference of the two expressions is at least -.5692 unit with 95% confidence.

11.3.2   Confidence Sets for Weighted Parameter Arrays

The preceding subsection discusses a weighted simultaneous confidence method, in which multiple parameters are treated as a vector of parameters (vectorization) with an application for the analysis of gene expression data. In the first look, it seems that one may simply vector the parameter array into a long vector and somehow adapt the weighted simultaneous confidence method of Yu et al (2022) [128] for the inference of parameter arrays. Unfortunately, as pointed out in Ma et al. (2023)[82] and Zhang et al (2017) [129], the intrinsic connection within each "subsystem" invalidates this vectorization approach in multiple learning. For example, in the study of treatment regime, once one of the medicines in the treatment regime is ineffective, the patient dies already, regardless of the significance of other components in the treatment regime (Thall et al. 2007 [117], Wang et al 2012 [123], or Wang et al. 2017 [122]). In conjunction with different weights associated with different "subsystems", a new methodology on weighted simultaneous confidence set for parameter arrays is on call.

This array-type inference problem can be viewed as a multivariate version of univariate simultaneous inference. Instead of rearranging all multiple parameters into one parameter vector, this subsection focuses on simultaneous inference for a parameter array formed by multiple parameter vectors, in which each parameter vector consists of multiple parameters. The extension from vector-wise multiple learning to array-wise multiple learning is analogical to the extension from univariate inference to multivariate statistical inference. We propose a stepwise confidence region method for inference on parameter arrays. More details on materials discussed in this section can be found in Ma et al. (2023)[82].

Let $H_i, i = 1, ..., k$ be k hypotheses with null hypothesis $H_{0i} : \theta_i \in \Theta_i$ versus alternative hypothesis $H_{1i} : \theta_i \in \Theta_i^c$, where $\theta_i = (\eta_{i1}, ..., \eta_{in_i})'$ is a vector of n_i parameters. Suppose for each hypothesis H_i, we have a test ϕ_i. Let c_i be the weight of hypothesis H_i and $\hat{P}_i$ the corresponding p-value.

Similar to previous subsection, define an S-value as a function of the weighted p-value as follows:

$$S_i = \frac{\hat{P}_i}{c_i},$$

and let $S_{(1)} \leq ... \leq S_{(k)}$ be the ordered S-values:

$$S_{(i)} = \frac{\hat{P}_i^*}{c_i^*},$$

where $\hat{P}_i^*$ and c_i^* are, respectively, the p-value and weight corresponding to the ordered S-value $S_{(i)}$ with the corresponding parameter θ_i^*, denoted as $\theta_{(i)}$ in the sequel.

Also, denote l_i, $i = 1, ..., k$, the dimension of the parameter space corresponding to $S_{(i)}$. For instance, if S_4 ranks the second among all the S-values, $S_{(2)}$, by the notation, we have $n_4 = l_2$. For notational convenience, define the cumulative weight

$$w_i = \sum_{j=k-i+1}^{k} c_j^*,$$

or equivalently,

$$w_{k-i+1} = \sum_{j=i}^{k} c_j^*.$$

In testing the multivariate parameter array $\theta_1, ..., \theta_k$, we focus on detecting the significance of the vector θ_i (indicating the function of the ith subsystem) in each hypothesis. For convenience, we confine our investigation to the case where the significance of the parameter vector is claimed when at least one of its components is significant (a scenario for the functioning of subsystems in a parallel design).

For a multiple testing array problem $H_0^i : \theta_i \in \Theta_i$ versus $H_1^i : \theta_i \in \Theta_i^c$, $i = 1, ..., k$, assume that there exist inverted confidence sets $D_1^*(Y), ..., D_k^*(Y)$, which are directed toward the alternative spaces $\Theta_1^c, ..., \Theta_k^c$, respectively. When screening up from $S_{(1)}$ to $S_{(k)}$, define m as the smallest integer such that $S_{(i)} \geq \frac{\alpha}{w_{k-i+1}}$, $1 \leq i \leq k$. Also, let $\Theta_{(i)}^c$ be the alternative space associated with the ith ordered weighted S-value $S_{(i)}$.

Theorem 11.6 *With the above setting, we have*

$$P(\theta \in L(\mathbf{Y}, m)) \geq 1 - \alpha, \text{ where}$$

$$L(\mathbf{Y}, m) = \begin{cases} D_{(1)}^*(Y) \times R^{l_2} \cdots \times R^{l_k}, \text{ when } m = 1 \\ \Theta_{(1)}^c \times \cdots \times \Theta_{(m-1)}^c \times D_{(m)}^*(Y) \times R^{l_{m+1}} \cdots \times R^{l_k}, \\ \qquad \text{when } 2 \leq m \leq k - 1 \\ \Theta_{(1)}^c \times \cdots \times \Theta_{(k)}^c, \text{ when } m = k. \end{cases}$$

As discussed in Ma et al. (2023), the above theorem leads to the following operating algorithm.

Algorithm: Weighted Confidence Procedure on Parameter Array

Step 1   If $S_{(1)} < \dfrac{\alpha}{w_k}$, then assert $\theta_{(1)} \in \Theta^c_{(1)}$ and go to Step 2 ;

else, assert $\theta_{(1)} \in \{\theta : \hat{P}^*_1 \geq \dfrac{c^*_1 \alpha}{w_k}\}$, and stop.

Step 2   If $S_{(2)} < \dfrac{\alpha}{w_{k-1}}$, then assert $\theta_{(2)} \in \Theta^c_{(2)}$ and go to Step 3 ;

else, assert $\theta_{(2)} \in \{\theta : \hat{P}^*_2 \geq \dfrac{c^*_2 \alpha}{w_{k-1}}\}$, and stop.

$\vdots$

Step k   If $S_{(k)} < \dfrac{\alpha}{w_1}$, then assert $\theta_{(k)} \in \Theta^c_{(k)}$ and stop;

else, assert $\theta_{(k)} \in \{\theta : \hat{P}^*_k \geq \alpha\}$, and stop.

The problem described above can be symbolically formulated as follows. Assume that we are interested in making inference for k parameter vectors θ_1, ..., θ_k simultaneously, where for the ith vector, $i = 1, ..., k$, the testing problem is

$$H_{0i} : \theta_i \in \Theta_i \quad versus \quad H_{1i} : \theta_i \in \Theta^c_i,$$

where

$$\Theta_i = \Theta_{i1} \times \cdots \times \Theta_{in_i} \in R^{n_i}.$$

It should be noted that this section is designed for simultaneous inference on parameter arrays. It is fundamentally different from previous subsection focusing on multiple learning of parameter vectors. The intrinsic connection within each parameter vector necessitates a new multivariate version of confidence region in this section. We shall now use a dataset to illustrate the use of the above simultaneous confidence method.

The dataset was originally discussed in Cortez et al (2009) [38], with two sub datasets containing wine information (quality index and physicochemical variables) for red and white variants, respectively, of the Portuguese "Vinho Verde" wine. It consists of the following physicochemical variables: fixed acidity; volatile acidity; citric acid; residual sugar; chlorides; free sulfur dioxide; total sulfur dioxide; density; pH-level; sulfates; and alcohol. It also contains the quality index based on the sensory of the wine samples.

Consider the intertwining relationships and the chemical nature of the 11 physicochemical variables, we partition the variables into three categories. Category I is about acidity, which includes "fixed-acidity", "volatile-acidity", "citric-acid" and "pH-level". Category II focuses on sulfur-dioxide, which includes "free sulfur-dioxide" and "total sulfur-dioxide". Category III contains all other chemical components which have great impact on the density of wine. This includes "residual-sugar", "chlorides", "sulfate", "alcohol", and "density" itself.

Example 11.12 *For the datasets on red wine and white wine described above, we are interested in learning wine features that significantly affect the wine quality index in tasting.*

Next, we used wine quality as a guideline to determine the standard/baseline group and the comparison groups. The baseline (denoted as group 0) is chosen to be wine variants with quality 3 or 4 ($n_{\mathrm{red}} = 63, n_{\mathrm{white}} = 183$), while four treatment groups are those with quality 5 (denoted as group 1, $n_{\mathrm{red}} = 679, n_{\mathrm{white}} = 1457$), 6 (denoted as group 2, $n_{\mathrm{red}} = 638, n_{\mathrm{white}} = 2195$), 7 (denoted as group 3, $n_{\mathrm{red}} = 199, n_{\mathrm{white}} = 880$), and more than 8 (denoted as group 4, $n_{\mathrm{red}} = 18, n_{\mathrm{white}} = 180$).

Since the p-values associated with Categories II and III are larger than the cutoff values. they are not statistically significant on wine quality, we only report the analysis result for Category I here. The inference problems of interest are as follows.

Denote M_i the median acidity level of group i for $i = 0, 1, 2, 3, 4$.

$$H_{10} : M_1 = M_0 \quad vs \quad H_{11} : M_1 > M_0$$

$$H_{20} : M_2 = M_0 \quad vs \quad H_{21} : M_2 > M_0$$

$$H_{30} : M_3 = M_0 \quad vs \quad H_{31} : M_3 > M_0$$

$$H_{40} : M_4 = M_0 \quad vs \quad H_{41} : M_4 > M_0$$

For instance, the first null hypothesis H_{10} states that the median acidity level of wine with quality score 5 equals the median acidity level of wine with baseline quality, and the alternative hypothesis H_{11} states that the median acidity level of wine with quality score 5 is higher than the median acidity level of wine with baseline quality.

For illustrating purposes, we consider two weight vectors for each category, equal weight $w_1 = (1, 1, 1, 1)$, and weights emphasizing high-quality wine $w_2 = (1, 1, 1, 100)$. More discussions on wine quality information can be found in Cortez et al (2009) [38].

The significance level in Table 11.3 is set to $\alpha = 0.005$. Denote M_i the median acidity level of group i for $i = 0, 1, 2, 3, 4$ and consider the following hypotheses,

$$H_{10} : M_1 = M_0 \quad vs \quad H_{11} : M_1 > M_0$$

$$H_{20} : M_2 = M_0 \quad vs \quad H_{21} : M_2 > M_0$$

$$H_{30} : M_3 = M_0 \quad vs \quad H_{31} : M_3 > M_0$$

$$H_{40} : M_4 = M_0 \quad vs \quad H_{41} : M_4 > M_0$$

For instance, the first null hypothesis H_{10} states that the median acidity level of wine with quality score 5 equals the median acidity level of wine with baseline quality, and the alternative hypothesis H_{11} states that the median acidity level of wine with quality score 5 is larger than the median acidity level of wine with baseline quality.

TABLE 11.3
Analytic results of Category I for red wine.

unweighted inference				
Quality	S-value	Weight	α^*	Conclusion
$Score = 7$	4.69E-5	1	0.0125	Claim H_{31}
$Score \geq 8$	0.02164	1	0.01667	Fail to reject H_{40}
$Score = 6$	0.02463	1	0.025	No conclusion
$Score = 5$	0.03974	1	0.05	No conclusion
weighted inference				
Quality	S-value	Weight	α^*	Conclusion
$Score = 7$	4.69E-5	1	0.000485	Claim H_{31}
$Score \geq 8$	2.164E-4	100	0.00049	Claim H_{41}
$Score = 6$	0.02463	1	0.025	Claim H_{21}
$Score = 5$	0.03974	1	0.05	Claim H_{11}

With the above setting, we run the proposed procedure on all three categories with different types of wines to investigate the impact of different groups of physicochemical variables on wine quality. Since the distribution of the data set is skewed, we use the Wilcoxon rank-sum test to compute the individual p-values for each test. As shown in Theorem 11.6, the procedure is valid for any test statistic suitable for the distribution of the data. The vector-wise p-value, S-value, adjusted significance level α^* are reported in Table 11.3.

Without weighting adjustment for red win, results summarized in Table 11.3 show the impact of acidity on the taste of red-wine. With equal weight, the algorithm stops at the second step, concluding that acidity is significant in differentiating moderately high-quality wine (Score = 7) from the baseline, but it is not significant for the rest. However, the inference result may be confounded by the unbalanced sample sizes between the group with score 7 ($n = 199$) and the group with scores higher than 8 ($n = 18$). Also, such inference conclusion is not coherent in terms of asserting the impact of acidity changes on the taste quality changes. It makes more sense to claim significant impact for wines with higher scores before claiming significant impact for wines with lower scores. Since we are interested in seeking factors that are important for high-quality wine, we consider a weight of 100 for the group in which the wine taste quality score is 8 and higher. With the weight vector of w_2, we are able to reach the conclusion that is consistent and coherent throughout the comparison groups. The new weighting scheme leads to the conclusion that acidity is significant in distinguishing red wine taste quality.

Now, consider the white wine Category I. Similar to the analysis for red wine, the impacts of Categories II and III are not statistically significant, we only report the analysis on Category-I here.

Comparing the results from unweighted and weighted procedures in Table 11.4, we can see that different weights may lead to different testing or-

TABLE 11.4

Analytic results of Category I for white wine.

unweighted inference				
Quality	S-value	Weight	α^*	Conclusion
$Score = 7$	2.24E-4	1	0.00125	Claim H_{31}
$Score = 6$	1.60E-3	1	0.00167	Caim H_{21}
$Score \geq 8$	2.84E-3	1	0.0025	Fail to reject H_{40}
$Score = 5$	7.59E-3	1	0.005	No conclusion
weighted inference				
Quality	S-value	Weight	α^*	Conclusion
$Score \geq 8$	2.84E-5	100	4.85E-5	Claim H_{41}
$Score = 7$	2.24E-4	1	0.00167	Claim H_{31}
$Score = 6$	1.60E-3	1	0.0025	Claim H_{21}
$Score = 5$	7.59E-3	1	0.005	Fail to reject H_{10}

der. For illustrating purpose, we chose a more precise significance level with $\alpha = 0.005$.

The equal weighting scheme results in an incoherent conclusion in which a relatively lower quality (Score = 7) wine is significant but higher quality (Score = 8 or more) wine is not significant. Comparatively, the weighted procedure in Table 11.4 leads to a more persuadable and interpretable assertion, that acidity is a significant factor in distinguishing white wine quality, because we have sufficient evidence to reject the null hypothesis.

Practically, the impact of acidity on wine taste is complex. However, industrial evidence shows that within a reasonable range relatively higher acidity level makes both red and white wine taste fresher than those with low acidity levels (see, for example, Plane et al 1980 [94]). Moreover, the sweetness might balance the taste of acidity and consequently the confounded taste may be more abundant. This agrees with the inference conclusion derived from the weighted confidence procedure, as stated in Table 11.3.

Inference on the above-mentioned settings calls for further developments of weighted inference for underlying parameter arrays. While we normally go with multiple testing procedures for the comparisons of more than one populations, it should be noted that the method of simultaneous confidence set plays a more effective role in estimating the unknown sets of parameters. There are distinct differences between the confidence set method and the multiple testing procedures (such as the closed testing principle). The closed testing principle essentially draws conclusions on multiple hypotheses by controlling the family-wise error rate. In the literature, various multiple testing procedures were proposed with different criteria in controlling the error rate, such as the control of false discovery rate, see for example, Hochberg and Tamhane (1987) [58] . The simultaneous confidence set, on the other hand, provides an overall estimation for all the parameters under investigation. Usually, a multiple testing procedure is unable to provide parameter estimation for follow-up investigations. However, simultaneous confidence sets are able to strongly control the overall error rate on testing the associated hypotheses. Compared with test-based inference methods, the method of simultaneous confidence set is more straightforward, interpretable, and reliable. More discussions on this regard for clinical trial studies can be found, for example, in papers of Chen (2008a)[20] and Chen (2008b) [21], or in the book of Chen (2014) [22].

SUMMARY Previous chapters basically deal with learning from a dataset representing a population. However, in real-life situations, we are often confronted with scenarios where the data for statistical learning and prediction come from multiple data sources or phases. This results in multiple error rates at different stage or data sources. Ignoring the intertwining relationship on the false positive or false negative rates may result in serious misleading conclusion partly due to accumulating error rates from different data learning resources. This chapter is thus motivated to synthesize published methodologies and algorithms handling multiple data resources.

We start with methods in sequential analysis where it is desirable to control both the probability of type-I error and the probability of type-II errors. We elucidate Wald's sequential likelihood ratio test for inference decisions, and the two-stage estimation approach for additional observations in the sequential sample. Sequential analysis appears in various learning phases such as for customer record data, medical history data, to list just a few.

Besides sequential data coming from multiple phases, more often we have data coming from multiple sources, or we are interested in learning multiple information at the same time. This directly relates to the vast literature on multiple comparisons and simultaneous inference. In this chapter, to give a glimpse into the field by introducing two key directions. The first one is on the prediction of dose-response curves in efficacy studies. This includes a step-up confidence procedure and the method of simultaneous confidence bands controls the overall coverage probability with proper adjustments on multiplicity. Further information can be found in papers [27], [26], and [105].

The second direction on recent development of simultaneous inference focuses on testing weighted hypotheses. This challenges conventional methods that implausibly assume equal weights for all population involved. It extends the conventional setting of equal weight inference into unequal weights according to related information on the populations. Two confidence methods are synthesized in this chapter. One is on weighted confidence regions that improves the prediction on genes significantly associated with breast cancer, by including biological information on gene expression data. The second approach directly extends the vector-wise multiple inference to array-wise confidence procedures as in Ma et al. (2023, [82]).

Since it is not plausible to always assume that all the hypothesis in a multiple testing problem shall be treated with the same weight, the two methods discussed in the last section of the book cast high lights on statistical learning for data with multiple resources.

Bibliography

[1] H. Ajhajala, M. Hendricks-Jones, J. Shawver, A. Amllay, J. T. Chen, and M. Hajjar et al. Expansion of telestroke coverage in community hospitals: Unifying stroke care and reducing transfer rate. *Annals of Neurology*, 12:1–7, 2023.

[2] R. Amorim and C. Hennig. Recovering the number of clusters in data sets with noise features using feature rescaling factors. *Information Sciences*, 324:126–145, 2015.

[3] R. Amorim and B. Mirkin. Minkowski metric, feature weighting and anomalous cluster initializing in K-means clustering. *Pattern Recognition*, 45:1061–1075, 2012.

[4] F. Aziz, J. Chen, and A. Comerota. Catheter-directed thrombolysis of iliofemoral deep vein thrombosis reduces deep vein thrombosis recurrence. *Journal of Vascular Surgery*, 53:252–254, 2011.

[5] G. Bell, T. Hey, and A. Szalay. COMPUTER SCIENCE: Beyond the data deluge. *Science*, 323:1297–1298, 2009.

[6] P. Bühlmann and B. Yu. Analyzing bagging. *The Annals of Statistics*, 30:927–961, 2008.

[7] P. Bühlmann and B. Yu. Boosting with the L2 loss: regression and classification. *Journal of the American Statistical Association*, 98:324–39, 2009.

[8] H. Block and J. T. Chen. Multivariate product-type lower bounds. *Journal of Applied Probability*, 38:407–420, 2001.

[9] L. Breiman. Bagging predictors. *Machine Learning Journal*, 26:123–140, 1996.

[10] L. Breiman. The heuristics of instability in model selection. *Ann. Statist.*, 24:2350–2381, 1996.

[11] L. Breiman. Arcing classifiers (Discussion paper). *Machine Learning Journal*, 26:801–824, 1998.

[12] L. Breiman. Random forests. *Machine Learning J.*, 45:5–32, 2001.

[13] L. Breiman. Statistical modeling: The two cultures. *Statistical Science*, 16(3):199–231, 2001.

[14] L. Breiman and J. Friedman. Estimating optimal transformations in multiple regression and correlation. *J Amer. Statist. Assoc.*, 80:580–619, 1985.

[15] D. Brent, M. Baugher, J. Bridge, J. T. Chen, and L. Chiappetta. Age and sex-related risk factors for adolescent suicide. *Journal of the American Academy of Child and Adolescent Psychiatry*, 38(12):1497–1505, 1999.

[16] G. Casella and R. Berger. *Statistical Inference*. Brooks/Cole Cengage Learning, Belmont, CA, USA, 2002.

[17] J. T. Chen. Optimal lower bounds for bivariate probabilities. *Advances in Applied Probability*, 30:476–492, 1998.

[18] J. T. Chen. Letter to the Editor: On the estimation of attributable risk in case-control studies. *Statistics in Medicine*, 20:979–982, 2001.

[19] J. T. Chen. A lower bound using Hamilton-type circuit and its applications. *Journal of Applied Probability*, 40:1121–1132, 2003.

[20] J. T. Chen. A two-stage estimation procedure. *Biometrics*, 64:406–412, 2008a.

[21] J. T. Chen. Inference on the minimum effective dose using binary data. *Communication in Statistics*, 38:2124–2135, 2008b.

[22] J. T. Chen. *Multivariate Bonferroni-Type Inequalities, Theory and Applications*. Chapman and Hall, CRC, Boca Raton, 2014.

[23] J. T. Chen. A nonparametric coherent confidence procedure. *Communications in Statistics – Theory and Methods*, 45:3397–3409, 2016.

[24] J. T. Chen and A. J. Comerota. Detecting the association between residual thrombus and post-thrombotic classification of chronic venous disease with range regression. *Reviews on Recent Clinical Trials*, 7:329–334, 2012.

[25] J. T. Chen, A. K. Gupta, and C. Troskie. Distribution of stock returns when the market is up (down). *Communications in Statistics-Theory and Methods*, 32:1541–1558, 2003.

[26] J. T. Chen and F. Hoppe. Simultaneous confidence intervals and regions. *Wiley StatsRef Statistics Reference Online ISBN 9781118445112*, John Wiley and Sons:1–10, 2017.

[27] J. T. Chen and F. M. Hoppe. Simultaneous confidence intervals. *The Encyclopedia of Biostatistics*, 5:4114–4116, 1998.

[28] J. T. Chen and F. M. Hoppe. A connection between successive comparisons and ranking procedures. *Statistics and Probability Letters*, 67:19–25, 2004.

[29] J. T. Chen, F. M. Hoppe, S. Iyengar, and D. Brent. A hybrid logistic regression model for case-control studies. *Methodology and Computing in Applied Probability*, 5:419–426, 2003.

[30] J. T. Chen, S. Iyengar, and D. Brent. Constraint estimation for the population attributable risk. *Journal of Applied Probability and Statistics*, 2:251–265, 2007.

[31] J. T. Chen and E. Seneta. A note on bivariate Dawson-Sankoff-type bounds. *Statistics and Probability Letters*, 24:99–104, 1995.

[32] J. T. Chen and E. Seneta. Multivariate Bonferroni-type lower bounds. *Journal of Applied Probability*, 33:729–740, 1996.

[33] J. T. Chen and E. Seneta. A refinement of multivariate Bonferroni-type inequalities. *Journal of Applied Probability*, 37:276–282, 2000.

[34] J. T. Chen and E. Seneta. A Fréchet-optimal strengthening of the Dawson-Sankoff lower bound. *Methodology and Computing in Applied Probability*, 8(2):255–264, 2006.

[35] J. T. Chen, E. Walsh, and A. Comerota et al. A new multiple test approach for nursing care administration of deep vein thrombosis patients. *Istanbul University Journal of the School of Business Administration*, 1:22–34, 2011.

[36] L. Chen and J. Chen. Refined Machine Learning Approaches for Mask Policy Analysis, Book chapter edited by E. Cetin and H. Ozen, Springer Nature, Singapore . *Health-care Policy, Innovation and Digitalization*, 10:197–211, 2024.

[37] A. J. Comerota, N. Grewal, J. T. Martinez, and J. T. Chen et al. Postthrombotic morbidity correlates with residual thrombus following catheter-directed thrombolysis for iliofemoral deep vein thrombosis. *Journal of Vascular Surgery*, 55:768 – 773, 2012.

[38] P. Cortez, A. Cerdeira, F. Almeida, T. Matos, and J. Reis. Modeling wine preferences by data mining from physicochemical properties. *Decision Support Systems*, 47:547–553, 2009.

[39] N. Cristianini and J. Shawe-Taylor. *An Introduction to Support Vector Machines*. Cambridge Univ. Press., 2000.

[40] C. Daniel and F. Wood. *Fitting Equations to Data*. Wiley, New York., 1971.

[41] A. Davision and D. Hinkley. *Bootstrap Methods and Their Applications.* Cambridge University Press, 2006.

[42] A. Dempster. Logicist statistic 1. Models and modeling. *Statistical Science*, 13:248–276, 1998.

[43] V. Dhar. Data science and prediction. *Communications of the ACM*, 12(56):64–73, 2013.

[44] P. Diaconis and B. Efron. Computer intensive methods in statistics. *Scientific American*, 248:116–130, 1983.

[45] C. Dunnett. A multiple comparison procedure for comparing several treatments with a control. *Journal of American Statistical Association*, 50:1096–1121, 1955.

[46] J. Eeckhoute, E. Keeton, and M. Lupien et al. Positive cross-regulatory loop ties GATA-3 to estrogen receptor α expression in breast cancer. *Cancer Research*, 67:6477–6483, 2007.

[47] B. Efron. Bootstrap methods, another look at the Jackknife. *The Annals of Statistics*, 7:1–26, 1979.

[48] B. Efron and R. Tibshirani. *An Introduction to the Bootstrap.* Chapman and Hall/CRC, 1993.

[49] E. Forgy. Cluster analysis of multivariate data: efficiency vs interpretability of classifications. *Biometrics*, 21:768–780, 1965.

[50] J. Galambos and I. Simonelli. Characterizations of probability distributions by properties of products of random variables. *Journal of Applied Statistical Science*, 13:1–10, 2004.

[51] I. Goodfellow, Y. Bengio, and A. Courville. *Deep Learning.* MIT press, 2016.

[52] A. Gupta and J. T. Chen. Goodness of fit test for the skew-normal distribution. *Communication in Statistics*, 30(4):907–930, 2001.

[53] A. Gupta and J. T. Chen. A class of multivariate skew-normal models. *The Annals of the Institute of Statistical Mathematics*, 56(2):305–315, 2004.

[54] M.H. Hansen and B. Yu. Model selection and the principle of minimum description length. *Journal of the American Statistical Association*, 96:746–774, 2002.

[55] J. Hartigan and M. Wong. A K-means clustering algorithm. *Journal of the Royal Statistical Society Series C: Applied Statistics*, 28:100–108, 1979.

[56] T. Hastie, R. Tibshirani, and Friedman J. *The Elements of Statistical Learning.* Springer, New York, 2009.

[57] I. Hedenfalk, D. Duggan, and Y. Chen et al. Gene-expression profiles in hereditary breast cancer. *Journal of Medicine*, 344:539–548, 2001.

[58] Y. Hochberg and A. C. Tamhane. *Multiple Comparison Procedures.* Wiley, New York, 1987.

[59] M. Hollander and D. Wolfe. *Nonparametric Statistical Methods.* John Wiley and Sons, Inc, New York, second edition, 1999.

[60] S. Holm. A simple sequentially rejective multiple test procedure. *Scandinavian Journal of Statistics*, 6:65–70, 1979.

[61] F. M. Hoppe. *Multiple Comparisons, Selection Procedures and Applications in Biometry.* Dekker, New York, 1993.

[62] F. M. Hoppe. Improving probability bounds by optimization over subsets. *Discrete Math.*, 306(5):526–530, 2006.

[63] F. M. Hoppe. The effect of redundancy on probability bounds. *Discrete Math.*, 309(1):123–127, 2009.

[64] F. M. Hoppe and E. Seneta. A Bonferroni-type identity and permutation bounds. *International Statistical Review*, 58:253–361, 1990.

[65] F. M. Hoppe and E. Seneta. Gumbel's identity, binomial moments, and Bonferroni sums. *International Statistical Review*, 80:269–292, 2012.

[66] H. Hotelling. Analysis of a complex of statistical variables into principal components. *Journal of Educational Psychology*, 24:417–441, 1933.

[67] J. C. Hsu. *Multiple Comparisons: Theory and Methods.* London, Chapman and Hall/CRC, 1996.

[68] J. C. Hsu and R. Berger. Stepwise confidence intervals without multiplicity adjustment for dose-response and toxicity studies. *Journal of American Statistical Association*, 94:468 – 482, 1999.

[69] Y. Huang, H. Xu, V. Calian, and J. Hsu. To permute or not to permute. *Bioinformatics*, 22:2244–2248, 2006.

[70] S. Iyengar. On a lower bound for the multivariate normal Mill's ratio. *Annals of Probability*, 4:1399–1403, 1986.

[71] S. Iyengar. Evaluation of normal probabilities of symmetric regions. *SIAM Journal on Scientific and Statistical Computing*, 3:418–423, 1988.

[72] G. James, D. Witten, T. Hastie, and R. Tibshirani. *An Introduction to Statistical Learning.* Springer-Science and Business Media, LLC, New York, 2013.

[73] E. Jauffret, F. Monville, F. Bertucci, and B. Esterni et al. Moesin Expression is A Marker of Basal Breast Carcinomas. *International Journal of Cancer*, 121:1779–1785, 2007.

[74] E. Kaizar, Y. Li, and J. Hsu. Permutation Multiple Tests of Binary Features Do Not Uniformly Control Error Rates. *Journal of the American Statistical Association*, 106:1067–1074, 2011.

[75] L. Kerns and J. T. Chen. A note on range regression analysis. *Journal of Applied Probability and Statistics*, 11:19–27, 2016.

[76] L. Kerns and J. T. Chen. Simultaneous confidence bands for restricted logistic regression models. *Journal of Applied Statistics*, 44:2036–2051, 2017.

[77] T. Lai. Sequential analysis: Some classical problems and new challenges. *Statistica Sinica*, 11:303–408, 2001.

[78] Y. LeCun, Y. Bottou, Y. Bengio, and P. Haffner. Gradient-based learning applied to document recognition. *Proceedings of the IEEE*, 86:2278–2324, 1998.

[79] C. Lee. *Self-Optimizing Networks*. Princeton Univesity senior thesis, 2017.

[80] E. Lehmann and G. Casella. *Theory of Point Estimation*. Springer-Verlag Inc, New York, 1998.

[81] E. Lehmann and J. Romano. *Testing Statistical Hypotheses*. Springer-Science and Business Media, LLC, 2005.

[82] Y. Ma, A. Yeh, and J. T. Chen. Simultaneous confidence regions and weighted hypotheses on parameter arrays. *Methodology and Computing in Applied Probability*, 25:1–18, 2023.

[83] J. MacQueen. Some methods for classification and analysis of multivariate observations. *Berkeley Symp. on Math. Statist. and Prob.*, 5:281–297, 1965.

[84] P. McCullagh and J. Nelder. *Generalized Linear Models*. Chapman and Hall, London, 1989.

[85] N. Meinshausen and B. Yu. Lasso-type recovery of sparse representations for high-dimensional data. *The Annals of Statistics*, 37:246–270, 2011.

[86] W. Meisel. *Computer-Oriented Approaches to Pattern Recognition*. Academic Press, New York., 1972.

[87] J. Mi and A. Sampson. A comparison of the Bonferroni and Scheffe bounds. *Journal Statistical Planning and Inference.*, 36(1):101–105, 1993.

[88] D. Michie, D. Spiegelhalter, and C. Taylor. *Machine Learning, Neural and Statistical Classification.* Ellis Horwood, New York, 1994.

[89] T. F. Móri and G. J. Székely. On the Erdos-Rényi generalization of the Borel-Cantelli lemma. *Studia Sci. Math. Hungar*, 18:173–182, 1983.

[90] T. F. Móri and G. J. Székely. A note on the background of several Bonferroni-Galambos-type inequality. *Journal of Applied Probability*, 22:836–843, 1985.

[91] T. T. Nguyen, J. T. Chen, A. K. Gupta, and K. T. Dinh. A proof of the conjecture on positive skewness of generalized inverse Gaussian distributions. *Biometrika*, 90:245–250, 2003.

[92] K. Pearson. On lines and planes of closest fit to systems of points in space. *Philosophical Magazine*, 2:559–572, 1901.

[93] W. Piegorsch and G. Casella. Confidence bands for logisic regression with restricted preditor variables. *Biometrics*, 44:739–750, 1988.

[94] R. Plane, L. Mattick, and L. Weirs. An acidity index for the taste of wines. *Am J Enol Citic*, 31:265–268, 1980.

[95] R. Redfern, J. Chen, and S. Sibrel. Effect of thermomechanical stimulation during vaccination on anxiety, pain, and satisfaction in pediatric patients: A randomized controlled trial. *Journal of Pediatric Nursing*, 38:1–7, 2018.

[96] R. Redfern, K. Fleming, R. March, N. Bobulski, M. Kuehne, J. T. Chen, and M. Moront M. Thrombelastography-directed transfusion in cardiac surgery: Impact on postoperative outcomes. *The Annals of Thoracic Surgery*, 107:1313–1318, 2019.

[97] R. Redfern, J. Micham, S. Seegert, and J. Chen. A buzzy approach to ease the discomfort of a needle stick – A prospective, randomized controlled trial. *Pain Management Nursing*, 20:164–169, 2019.

[98] R. Redfern, J. Micham, D. Sievert, and J. Chen. Effects of thermomechanical stimulation during intravenous catheter insertion in adults: A prospective randomized study. *Journal of Infusion Nursing*, 41:294–300, 2018.

[99] R. Redfern, C. Rueta, S. O'Drobinak, J. Chen, and K. Beer. Closed incision negative pressure therapy effects on postoperative infection and surgical site complication after total hip and knee arthroplasty. *Journal of Arthroplasty*, 32:3333–3339, 2018.

[100] E. Seneta. On a genetic inequality. *Biometrics*, 29:810–814, 1973.

[101] E. Seneta. An inequality from genetics. *Advances in Applied Probability*, 18(3):860–861, 1986.

[102] E. Seneta. On the history of the strong law of large numbers and Boole's inequality. *Historia Mathematica*, 19(1):24–39, 1992.

[103] E. Seneta. Probability inequalities and Dunnett's test. Book chapter edited by F. M. Hoppe, Dekker, New York. *Multiple Comparisons, Selection, and Applications in Biometry*, 134:29–35, 1993.

[104] E. Seneta and J. T. Chen. Frechét optimality of upper bivariate Bonferroni-type bounds. *Theory of Probability and Mathematical Statistics*, 52:147–152, 1996.

[105] E. Seneta and J. T. Chen. A sequentially rejective test procedure. *Theory of Stochastic Processes*, 3(19)(3-4):393–402, 1997.

[106] E. Seneta and J. T. Chen. Multivariate Sobel-Uppuluri-Galambos type bounds. *Ukrainian Mathematical Journal*, 52(9):1283–1293, 2000.

[107] E. Seneta and J. T. Chen. On explicit and Fréchet optimal lower bounds. *Journal of Applied Probability*, 39:81–90, 2002.

[108] E. Seneta and J. T. Chen. A generator for explicit univariate lower bounds. *Statistics and Probability Letters.*, 75:256–266, 2005.

[109] E. Seneta and J. T. Chen. Simple stepwise tests of hypotheses and multiple comparisons. *International Statistical Review*, 73(1):21–34, 2005.

[110] J. Shao. *Mathematical Statistics*. Springer. 2nd edition., 2003.

[111] R. Shermis, R. Redfern, J. Bazydlo, G. Naimy, H. Kudrolli, and J. T. Chen. Performance of molecular breast imaging as an adjunct diagnostic tool. *University of Toledo Journal of Medical Sciences*, 6:15–19, 2019.

[112] P. Sprent and N. Smeeton. *Applied Nonparametric Statistical Methods, 3rd edition.* Chapman & Hall/CRC, 2001.

[113] M. Stone. Cross-validatory choice and assessment of statistical predictions. *J Roy. Statist. Soc. B*, 36:111–147, 1974.

[114] J. Storey and R. Tibshirani. Statistical Significance for Genomewide Studies. *Proceedings of the Nattional Acadamy of Science*, 27:4914–4927, 2003.

[115] A. Tamhane and L. Liu. On Weighted Hochberg Procedures. *Biometrika*, 95:279–294, 2008.

[116] J. Tank, G. Georgiadis, and J. Bair et al. Does the use of ethyl chloride improve patient-reported pain scores with negative pressure wound therapy dressing changes? A prospective, randomized controlled trial. *Journal of Trauma and Acute Care Surgery*, 6:1061–1066, 2021.

[117] P. Thall, C. Logothetis, L. Pagliaro, S. Wen, M. Brown, D. Williams, and R. Millikan. Adaptive therapy for androgen independent prostate cancer: A randomized selection trial including four regimens. *Journal of the National Cancer Institute*, 99:1613–1622, 2007.

[118] J. Tukey. The future of data analysis. *The Annals of Mathematical Statistics*, 33:1–67, 1962.

[119] V. Vapnik. *The Nature of Statistical Learning Theory*. Springer, New York, 1995.

[120] V. Vapnik. *Statistical Learning Theory*. Wiley, New York, 1998.

[121] G. Wahba. *Spline Models for Observational Data*. SIAM, Philadelphia, 1990.

[122] L. Wang, Y. Lin, and J. Chen. Simultaneous inference for treatment regimes. *Communications in Statistics – Theory and Methods*, 46:9679–9690, 2017.

[123] L. Wang, A. Rotnitzky, X. Lin, R. Millikan, and P. Thall. Evaluation of viable dynamic treatment regimes in a sequentially randomized trial of advanced prostate cancer. *Journal of the American Statistical Association*, 46:493–508, 2012.

[124] F. Weaver, A. Comerota, M. Youngblood, and J. Froechich. Surgical revascularization versus thrombolysis for nonembolic lower extremity native artery occlusions: Results of a prospective randomized trial. *Journal of Vascular Surgery*, 24:513–523, 1996.

[125] C. Wu. Future directions of statistical research in China: A historical perspective. *Application of Statistics and Management*, 1:1–7, 1986.

[126] P. Xie and T. Chen. A new proof of the positive definiteness on the sample covariance matrix. *Acta Sci. Natur. Univ. Sunyatseni*, 27:113 – 114, 1988.

[127] P. Xie and T. Chen. Positive definiteness of covariance matrices of continuous sample. *Acta Sci. Natur. Univ. Sunyatseni*, 29:102 – 104, 1990.

[128] Y. Yu, J.T. Chen, and B. Yeh. Weighted step-down confidence procedures with applications to gene expression data. *Communications in Statistics – Theory and Methods*, 51:2343–2356, 2022.

[129] Y. Zhang, J. T. Chen, S. Wang, C. Andrews, and A. Levy. How do consumers use nutrition labels on food products in the United States? *Topics in Clinical Nutrition*, 32:161–171, 2017.

[130] P. Zhao and B. Yu. On model selection consistency of Lasso. *The Journal of Machine Learning Research*, 7:2541–2563, 2006.

Index

annual revenue, 185
Aspirin efficacy, 262

Bonferroni summation, 196–199
bootstrapping, 24, 46–48, 50, 51, 53, 54
breast cancer study, 272
bun-to-creatinine, 73, 74, 79

cardiovascular safety score, 262
case-control, 3, 27, 28, 54
classification trees, 23, 30, 207, 214, 227
cohort study, 25–28
cross-sectional data, 25, 54
cubic spline, 163

data science, 11, 13, 14, 18, 22, 23, 28, 54, 55, 59, 96, 97, 100, 118, 148
data-driven, 1, 3, 5, 8, 11–14, 16, 18, 22, 24, 97
deep vain thrombosis, 270
diabetes, 2, 19, 20, 25, 65, 66, 174, 177, 238
directed confidence set, 262, 263
dose-response, 151, 261, 266, 284
duality theorem, 98, 185, 192, 202

entropy, 140, 207, 213, 217, 228
expected prediction error, 98

gene expression data, 272
Gini index, 210, 212, 215, 216, 218, 228

homogeneity measure, 206, 211, 215, 227

hyperplane, 187–189

insurance premium, 3–6, 11, 22, 121, 160
inverted confidence set, 263
iteration process, 200–202

jackknife, 38–40

K-means clustering, 229, 233, 235, 238

LASSO, 154, 156, 157, 165
LDL-cholesterol, 61, 64, 67, 68, 72, 85
leverage statistic, 147, 148
linear programming, 98, 185, 196, 199, 203
linear programming bound, 199, 200
linear programming lower bound, 199, 200
linear regression, 1, 4, 8, 34, 41, 46, 120, 123, 127, 128, 133, 139, 141, 146, 150, 155, 156, 160, 183
location invariant diagnostic predictor, 94
logistic regression, 2, 3, 14, 35, 58, 168, 176, 178, 182, 267, 268
LOOCV, 38, 39, 41, 44, 55

marketing analysis, 31
minimum risk classification, 167
minimum risk classifier, 171, 172, 177
minimum risk equivariant estimator, 110, 112, 118
minimum risk estimator, 168, 179, 234

Printed in the United States
by Baker & Taylor Publisher Services